ELECTRICAL CONTROL FOR MACHINES

Fifth Edition

STUDENT MANUAL

KENNETH B. REXFORD

Delmar Publishers

an International Thomson Publishing company I(T)P®

Albany • Bonn • Boston • Cincinnati • Detroit • London • Madrid
Melbourne • Mexico City • New York • Pacific Grove • Paris • San Francisco
Singapore • Tokyo • Toronto • Washington

Notice to the Reader

Publisher does not warrant or guarantee any of the products described herein or perform any independent analysis in connection with any of the product information contained herein. Publisher does not assume, and expressly disclaims, any obligation to obtain and include information other than that provided to it by the manufacturer.

The reader is expressly warned to consider and adopt all safety precautions that might be indicated by the activities described herein and to avoid all potential hazards. By following the instructions contained herein, the reader willingly assumes all risks in connection with such instructions.

The publisher makes no representations or warranties of any kind, including but not limited to, the warranties of fitness for particular purpose or merchantability, nor are any such representations implied with respect to the material set forth herein, and the publisher takes no responsibility with respect to such material. The publisher shall not be liable for any special, consequential or exemplary damages resulting, in whole or in part, from the readers' use of, or reliance upon, this material.

Delmar Staff

Publisher: Robert D. Lynch
Acquisitions Editor: Mark Huth
Developmental Editor: Michelle Ruelos Cannistraci
Production Coordinator: Toni Bolognino

Copyright © 1997
By Delmar Publishers
A division of International Thomson Publishing Inc.

The ITP logo is a trademark under license.

Printed in the United States of America

For more information, contact:

Online Services

Delmar Online
To access a wide variety of Delmar products and services on the World Wide Web, point your browser to:
http://www.delmar.com/delmar.html
or email: info@delmar.com

thomson.com
To access International Thomson Publishing's home site for information on more than 34 publishers and 20,000 products, point your browser to:
http://www.thomson.com
or email: findit@kiosk.thomson.com

A service of I(T)P®

Delmar Publishers
3 Columbia Circle, Box 15015
Albany, New York 12212-5015

International Thomson Publishing Europe
Berkshire House 168-173
High Holborn
London WC1V7AA
England

Thomas Nelson Australia
102 Dodds Street
South Melbourne, 3205
Victoria, Australia

Nelson Canada
1120 Birchmont Road
Scarborough, Ontario
Canada M1K 5G4

International Thomson Editores
Campos Eliseos 385, Piso 7
Col Polanco
11560 Mexico D F Mexico

International Thomson Publishing GmbH
Königswinterer Strasse 418
53227 Bonn
Germany

International Thomson Publishing Asia
221 Henderson Road
#05-10 Henderson Building
Singapore 0315

International Thomson Publishing — Japan
Hirakawacho Kyowa Building, 3F
2-2-1 Hirakawacho
Chiyoda-ku, Tokyo 102
Japan

5 6 7 8 9 10 XXX 02 01 00 99 98

Library of Congress Catalog Number: 91-14664
ISBN: 0–8273–8162X

Contents

Foreword

This manual was prepared to assist the student of Electrical Control for Machines in obtaining a more thorough knowledge of the material. The questions in this manual follow the text content in sequence from Chapter 1 through Chapter 19.

Most of the questions can be answered directly from the text by careful and complete reading. In some cases, the student is asked to express his or her own thoughts and ideas on the subject.

Many circuits are shown for correction or modification according to instructions. A few circuit design problems are included. These problems will be similar to those presented and explained in the text. It is suggested that the student use graph paper when drawing these circuits.

When using this manual, answer each question in the spaces provided. Circuits should be drawn on a separate sheet of graph paper as suggested. To "learn by doing" is one of the more effective methods of education.

Chapter 1
Transformers and Power Supplies

1. The turns ratio on a given control transformer is 2:1. If a voltage of 480 volts is connected to the primary of the transformer, what is the secondary voltage?

2. Errors have been made in the drawing shown in Figure 1-1. Use graph paper to correctly draw this figure.

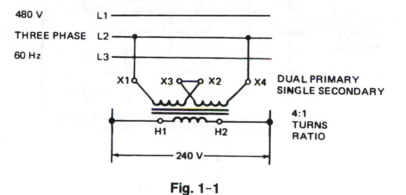

Fig. 1-1

3. What is meant by regulation in a transformer?

4. You plan to use an open-type transformer mounted in an enclosure with other control devices. What problem can develop?

5. What is meant by a step-down transformer?

6. Using the regulation curve shown in Figure 1–2, what is the approximate percent of output voltage when there is a 50-ampere secondary load?

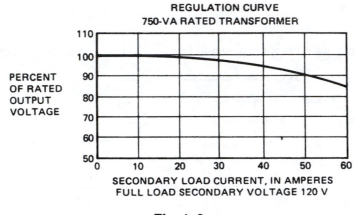

REGULATION CURVE
750-VA RATED TRANSFORMER

PERCENT OF RATED OUTPUT VOLTAGE

SECONDARY LOAD CURRENT, IN AMPERES
FULL LOAD SECONDARY VOLTAGE 120 V

Fig. 1-2

7. In reviewing the electrical load on a given machine, consisting of relays, contactors, timers, and motor starters, you find that the total inrush current is 14 amperes. The total sealed current is 3.5 amperes. The control voltage is 120. Using the second method shown in the text, calculate the size control transformer you would use for this job.

8. What causes temperature rise in a transformer?

9. You have a 240-volt, three-phase, 60-hertz power line. Use graph paper and draw the symbols for the power line and a dual-primary, single-secondary control transformer (240-480/120 volt). Connect the transformer to one phase of the power line to supply 120 volts to the secondary.

10. What do you mean by a dual-secondary winding on a control transformer? Use graph paper and draw the symbol for a dual-primary, dual-secondary transformer showing the connections required to supply 120 volts on the secondary from 480 volts on the primary. Are transformers available with more than two voltages available on the primary? If so, list a few of the voltages available.

11. What control circuit application information does the transformer regulation curve supply?

12. You are using a 4:1 ratio control transformer. The line voltage drops from 480 to 420. What will be the resulting control voltage?

13. A transformer has a no-load voltage of 130 and a full-load voltage of 126. What is the percent regulation for this transformer?

14. Will a 5 kVA transformer designed to operate at 50 Hz operate satisfactorily at 60 Hz? Explain.

15. Two transformers are to be operated in parallel. One transformer has an impedance of 4%. The other transformer has an impedance of 5%. Will they operate satisfactorily when connected in parallel? Explain.

16. With many of the currently designed control transformers, what problem can exist if the transformer is mounted within a control cabinet?

17. In the sketch shown in Figure 1–3, will the connection provide a balanced load across a 3-phase line?

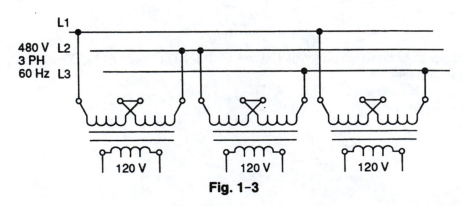

Fig. 1–3

18. You have a transformer that has been subjected to excessive temperatures. What can be the results?

19. With the CVR shown in the textbook, what is the function of a second pair of magnetic shunts and filter winding?

20. In the UPS shown in the textbook, what is the function of the microprocessor controlled converter?

21. In using the UPS as shown in the textbook, you are using a back-up battery. How will you know when the battery charge has been exhausted?

22. In the UPS shown in the textbook, for what amount of time will the standby battery maintain the unit's fully rated load?

TRUE AND FALSE STATEMENTS FOR CHAPTER 1

On the line provided, write TRUE if the statement is true or FALSE if the statement is false.

_____   1. Turns ratio in a transformer refers to the fact that voltage is indirectly proportional to the number of turns on each coil.

_____   2. The primary winding of a transformer carries the line voltage and the secondary carries the load voltage.

_____   3. The two primary coils in Figure 1–4 are connected in series.

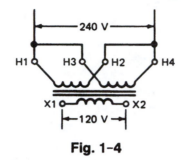

Fig. 1-4

_____   4. Voltage regulation in transformers is the difference between no-load voltage and full-load voltage expressed in percent.

_____   5. A step-up transformer with a 1:2 turns ratio has a greater output voltage than the input voltage.

_____   6. Special transformer primary windings are available with multiple taps for 200-208-240-480-575 volts.

_____   7. Given the transformer connections as shown in Figure 1–5, the secondary voltage will be 120.

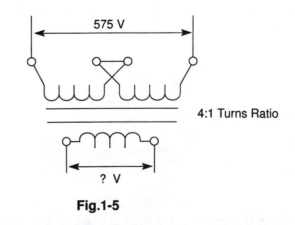

Fig.1-5

_____   8. It is generally desirable for a control transformer to have a regulation in excess of 10%.

_____   9. Two transformers with different voltage ratings are connected in parallel. This type of connection will have no affect on the life of the transformer.

_____ 10. When electrical power is generated, it generally is both clean and stable.

Chapter 2
Fuses, Disconnect Switches, and Circuit Breakers

1. How is protection provided on the disconnect switch and circuit interrupter?

2. Describe the standard one-time fuse and how it operates.

3. State two applications for the current-limiting fuse.

4. How can you determine the proper interrupting capacity needed for a protective device?

5. Use graph paper and draw a complete diagram using a 480-volt, three-phase power line, a three-pole thermalmagnetic circuit breaker, dual-primary, single-secondary control transformer with secondary protection.

6. Describe the mechanical operator option available for fused disconnect switches and circuit breakers mounted on a panel in a control enclosure.

7. Explain the inverse time characteristic of a fuse. Why is this characteristic desirable?

8. What is a rejection-type fuse?

9. The interrupting capacity of a rejection-type fuse is:
    a. 50,000 amperes.
    b. 100,000 amperes.
    c. 200,000 amperes.

10. From the fuse curve in Figure 2–1, what is the approximate melting time for the 100-ampere fuse at 2000 amperes?

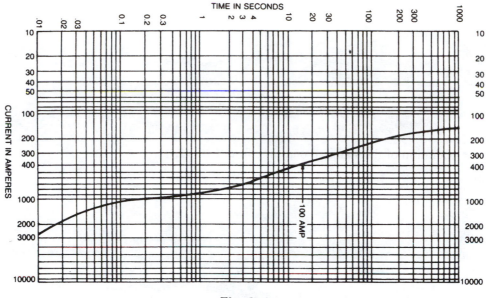

Fig. 2–1

11. A given plant has a 480-volt, three-phase, 60-hertz power line. A dual primary transformer is to be connected across one phase of the power line. The transformer is to deliver 120 volts to a control circuit. The power line disconnecting means and protection is supplied by a three-pole, fused disconnect switch. The transformer primary is protected by two fuses, one in each line. Use graph paper to draw a circuit incorporating these conditions.

12. When you are installing or replacing a protective device, what factor or factors should you consider?
    a. Voltage rating
    b. Quantity of fuses available
    c. Current rating
    d. Frequency rating
    e. Interrupting capacity

13. What information does the published interrupting capacity of a protective device supply?

14. Explain the difference between a standard fuse and a rejector type (Class R) fuse.

15. Describe the operation of the standard one-time fuse (Class K-5).

16. What do you mean by the time-delay characteristic of an efficient fuse?

17. What are some of the problems in protecting solid-state devices?

18. What is let-thru current as applied to a fuse?

19. When calculating the peak available current, why is it 2.3 times greater than the RMS available current?

20. Can you experience lightning voltage surges other than by a direct hit? Explain and compare.

21. How can you find the impedance of a transformer and how is it expressed? How is this information used?

22. What two factors determine the I^2t passed by a given fuse?

TRUE AND FALSE STATEMENTS FOR CHAPTER 2

On the line provided, write TRUE if the statement is true or FALSE if the statement is false.

_____   1. A fuse has a direct time current characteristic; i.e., the greater the overcurrent, the more time is required for the fuse to open a circuit.

_____   2. Class R fuses (rejection form) require the use of a rejection-type fuse block.

_____   3. The time-delay fuse is the best selection for use in protecting an induction motor.

_____   4. The current-limiting action of a fuse limits the thermal and mechanical stresses created by fault currents.

_____   5. The I^2t rating of a fuse refers to a particular manufacturer of the fuse.

_____   6. The t in the term I^2t is the total clearing time for the fuse.

_____   7. Surge voltages caused by induction on a line from lightning will result in the maximum damage.

_____   8. The combination of both magnetic and thermal trip units on circuit breakers is desirable.

_____   9. The DC resistance of a transformer is used in calculating the interrupting capacity for protective devices.

_____   10. Ambient temperature in the location where the protective device is located is important.

Chapter 3
Control Units for Switching
and Communication

1. What is the control rating for a typical heavy-duty, oil-tight unit?

2. Explain how the maintained contact attachment operates.

3. Use graph paper to draw the symbol for a three-position, four-contact selector switch. Contact 1 is closed in position 1, contact 2 is closed in position 2, and contacts 3 and 4 are closed in position 3.

4. Draw the symbol for a normally open foot-operated switch.

5. Explain some of the advantages for the push-to-test pilot light.

6. Explain some of the differences between a selector switch and a drum switch.

7. Errors have been made in the circuit drawing shown in Figure 3–1. Use graph paper to redraw the circuit making the proper corrections.

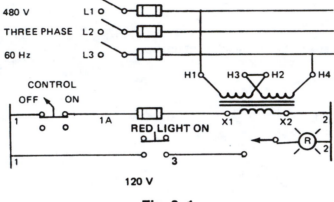

Fig. 3-1

8. Use graph paper to draw a circuit using a three-position selector switch and four pilot lights: one red, one green, one white, and one amber. The red light is to be energized in position 1. The green light is to be energized in position 2. The amber and white lights are to be energized in position 3.

9. Use graph paper to draw the complete electrical circuit for the push-to-test pilot light and explain how it operates.

10. You have a 480-volt, three-phase, 60-hertz power line. Disconnecting means and protection are provided by a three-pole thermal magnetic circuit breaker. A dual primary, single secondary control transformer provides 120 volts, single phase for a control circuit. The transformer primary is protected by two fuses. The secondary uses one fuse, with the other side grounded. A two-pole selector switch provides disconnecting means for the control circuit. When the selector switch is turned to the ON position, a push-to-test green pilot light is energized. This indicates the presence of control voltage. Use graph paper to draw a circuit to satisfy these conditions.

11. In the machine tool industry, the yellow push-button color is assigned to:

    a. stop-emergency stop.

    b. return-emergency return.

    c. start-motor cycle.

12. Errors have been made in drawing the circuit shown in Figure 3–2. Use graph paper to redraw the circuit, showing the necessary corrections. Using a three-position selector switch:

    The red light is energized in position 1.

    The white light is energized in position 3.

    The green light is energized in position 3.

    The amber light is energized in position 1.

    All lights are deenergized in position 2—control voltage 120.

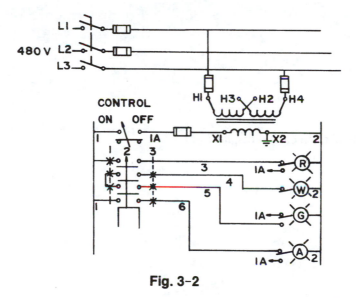

Fig. 3–2

13. Using one maintained contact push-button unit, two momentary contact mushroom-head push-button units, one green push-to-test pilot light, and one red push-to-test pilot light, design a circuit to accomplish the following:

    Operating the maintained contact push-button unit will energize the green push-to-test pilot light. After the green push-to-test pilot light is energized, the red push-to-test pilot light can be energized by operating both momentary push-button units.

    Use graph paper to draw the circuit.

14. Use graph paper to draw a circuit showing a four-position selector switch, three pilot lights (one red, one green, and one yellow) and three push-button switches. In position one, by operating push-button switch A, the red pilot light will be energized. In position 2, by operating push-button switch B, the green pilot light will be energized. In position 3, by operating push-button C, the yellow pilot light will be energized. In position 4, none of the pilot lights can be energized. Use push-to-test pilot lights. To complete the circuit, show a three-phase power line protected by a thermal magnetic circuit breaker, a control transformer with the primary side protected by two fuses, the secondary side protected by one fuse with the common side grounded. A control On/Off selector switch is used to switch the control power. Line voltage is 480 and control voltage is 120. Assign all appropriate control and power circuit numbers.

15. Where would you be likely to find the use of semigraphic displays using annunciators?

16. When the construction of a membrane switch allows it to be vented, what are some of the problems that may be encountered?

17. What is "birefringence" in crystal display?

18. What is one advantage of the dot-matrix type of liquid crystal display?

19. Use graph paper to draw the circuit for an LED and explain the purpose for the resistor.

20. In the circuit shown in Figure 3–3, do you find any errors and/or omissions? Explain.

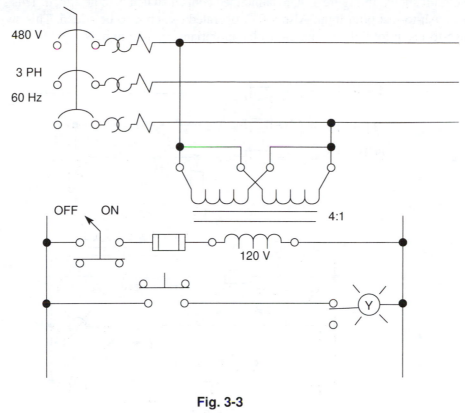

Fig. 3-3

21. Use graph paper to draw the symbol for a four-position, four-contact selector switch. Contact 1 is closed in position 1, contact 2 is closed in position 2, contact 3 is closed in position 3, and contact 4 is closed in position 4.

22. In the corrected circuit for Figure 3–3, add the selector switch called for in question 21. In position 1, a red push-to-test pilot light is to be energized. In position 2, a green push-to-test pilot light is to be energized. In position 3, an amber push-to-test pilot light is to be energized. In position 4, a white push-to-test pilot light is to be energized. Complete with all appropriate circuit numbers.

23. In the circuit shown in Figure 3–4, a maintained contact switch is to be added. This switch is to energize a green push-to-test pilot light. Also a foot-operated switch is to be added. This switch is to energize a red push-to-test pilot light. Complete with appropriate circuit numbers.

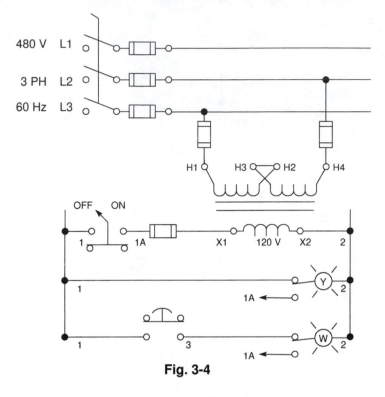

Fig. 3-4

TRUE AND FALSE STATEMENTS FOR CHAPTER 3

On the line provided, write TRUE if the statement is true or FALSE if the statement is false.

_____ 1. Oil-tight units can be either base mounted or panel mounted.

_____ 2. A selector switch with two contacts arranged for two positions may be called a double-pole, double-throw selector switch.

_____ 3. The symbol for a foot-operated switch is the same as the push-button switch except it has the letter F placed above the symbol.

_____ 4. Miniature oil-tight units may be used where space is at a premium.

_____ 5. When using a push-to-test pilot light, it is generally advisable to remove the bulb from the unit before testing.

_____ 6. The machine tool industry has never assigned specific colors to push-button units or pilot lights.

_____ 7. Lamp displays in annunciators are available in only one type of construction.

_____ 8. A resistor unit in a LED prevents the current from exceeding the maximum current rating.

_____ 9. To obtain maximum viewing angle in liquid crystal display presents no problem for the designer.

_____ 10. In the design of liquid crystal displays, a full-color display can be constructed by using color filters.

Chapter 4
Relays

1. What is the condition of a normally closed relay contact when the relay coil is energized?

2. What percent of rated voltage do manufacturers use as the minimum drop-out and pick-up voltage?

3. What advantage is gained in the use of gold-plated relay contacts?

4. What is meant by a split or bifurcated contact? What advantage is gained through its use?

5. Use graph paper to draw the symbols for the following time-delay relay contacts.

   a. Normally open time delay after energizing

   b. Normally closed time delay after deenergizing

6. Errors have been made in the circuit shown in Figure 4–1. Use graph paper to redraw, making the proper corrections.

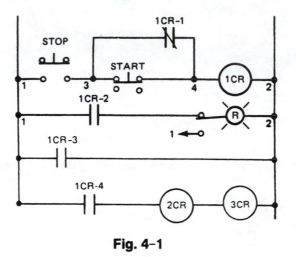

Fig. 4–1

7. Under what conditions would you consider the use of plug-in relays?

8. What are some of the uses for contactors?

9. Use graph paper to draw the complete power and control circuit using the following equipment and conditions.

   a. The power line is 480 volts, three-phase, 60-hertz.

   b. Protection and disconnecting means in the power line are provided through the use of a three-pole, fused disconnect switch.

   c. The control voltage is 120 volts, which is obtained through the use of a dual primary, isolated single secondary control transformer. Supply the transformer secondary protection and disconnecting means.

   d. The power load, energized at line voltage, consists of two separate banks of three heating elements each. They are connected in three-phase delta. The two banks are identified as #1 and #2.

   e. Two contactors are used. Each contactor has three power contacts and two auxiliary contacts.

   f. A red indicating light is used to show when #1 bank is energized. An amber light is used to show when #2 bank is energized.

   g. Operating #1 heat bank ON push-button switch energizes #1 heat bank. Operating #1 heat bank OFF push-button switch deenergizes #1 heat bank. Operating #2 heat bank ON push-button switch energizes #2 heat bank. Operating #2 heat OFF push-button switch deenergizes #2 heat bank.

10. Using the circuit shown in Figure 4–2, show how the circuit numbering system is used as you move from one side of the control circuit to the other. Assume you move through several components such as push-button switches, limit switches, pressure switches, and a relay coil.

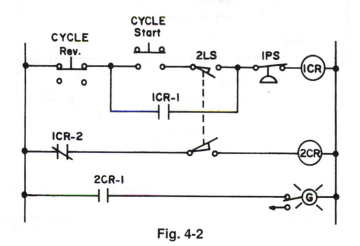

Fig. 4-2

11. Explain how the mercury-to-metal contactor operates.

12. See Figure 4–3. The sequence of operations proceeds as follows:

    a. Operate the START push-button switch.

    b. Relay coil 1CR is energized.

    Relay contact 1CR-1 closes, interlocking around the START push-button switch.

    Relay contact 1CR-2 closes, energizing the time-delay relay coil 1TR.

    Relay contact 1CR-3 closes, energizing the white pilot light.

    Do you find any problems with this sequence? Will the WHITE pilot remain energized until the STOP push-button switch is operated? Explain.

    (TO ONE PHASE OF THREE-PHASE POWER LINE.)

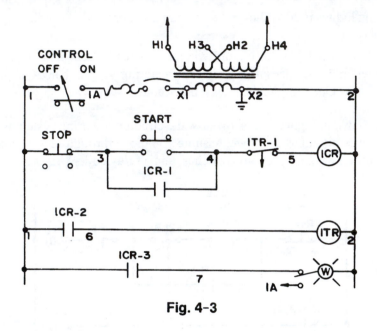

Fig. 4-3

13. What is the difference between pick-up voltage and drop-out voltage for a relay coil?

14. Give examples of an inductive load and a noninductive load.

15. In Figure 4–4, when will the green pilot light be energized?

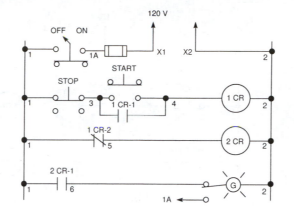

Fig. 4-4

16. Now change the N.C. 1CR-2 relay contact in Figure 4–4 to a normally open contact. What are the results now in reference to the energizing of the green pilot light?

17. Explain when the green and red pilot lights will be energized in the circuit shown in Figure 4–5.

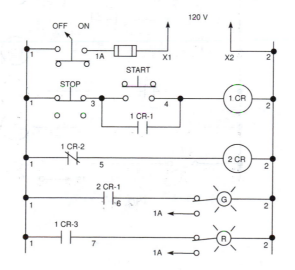

Fig. 4-5

18. Explain the operation of the circuit shown in Figure 4–6.

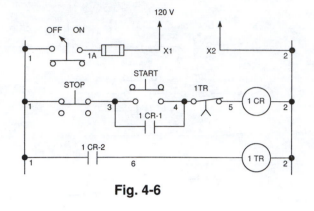

Fig. 4-6

19. What action occurs in the circuit shown in Figure 4–7 if the time on time-delay relay 1TR is set for less than the time set on time-delay relay 2TR?

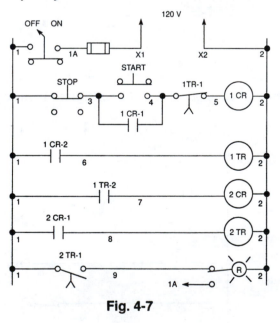

Fig. 4-7

20. Under what conditions can you use a contactor as a motor starter?

21. Explain the operation of the mercury contactor.

22. Why do you think the latching relay is sometimes referred to as a memory relay?

TRUE AND FALSE STATEMENTS FOR CHAPTER 4

On the line provided, write TRUE if the statement is true or FALSE if the statement is false.

_____   1. Due to the open gap in the magnetic path (circuit) in a relay, the initial current will be low.

_____   2. A relay coil will not pick up (energize) unless 100% of rated voltage is applied.

_____   3. With split or bifurcated contacts, the contact is divided into two parts. This provides for double the contact-make points.

_____   4. The symbols for the time-delay relay timing contacts are the same for both the N.O. and N.C. contacts.

_____   5. A contactor can be used as a motor starter if the proper overload relays are added.

_____   6. Auxiliary contacts are not available for most contactors.

_____   7. The basic use for the mercury contactor is the switching of resistance heating loads.

_____   8. The interlock circuit is almost never used except in the case of emergency.

_____   9. The symbol for the instantaneous normally closed contact is the same as the normally closed timing contact except for the direction of the arrow.

_____   10. The plug-in relay is available for both ac and dc. Mountings can be obtained with a tube-type socket, square-base socket, or flange mounting using slip-on connectors.

Chapter 5
Solenoids

1. What design features are improved with the oil-immersed solenoid?

2. In the double-solenoid operating valve, what keeps the valve spool in a center position when both solenoids are deenergized?

3. In the control circuit shown in Figure 5–1, errors have been made. Use graph paper to redraw the circuit, making the proper corrections.

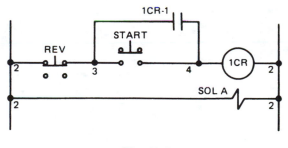

Fig. 5–1

4. Refer to the circuit you have corrected in question 3, Figure 5–1. Use graph paper to draw a new circuit adding a second solenoid valve B. The added solenoid valve is to be energized at the same time as solenoid A. Assume additional normally open contacts are available on relay 1CR.

5. Referring to the circuit you drew in question 4, it is now required to have an adjustable delay for the energizing of solenoid B after the energizing of solenoid A. Use graph paper to draw a new circuit to satisfy these requirements.

6. Use graph paper and design a control circuit using the components given below and accomplish the sequence of operations as given.

   a. Use a dual-primary, single-secondary control transformer.

   b. Connect the primary of the transformer to a single phase of a three-phase, 480-volt, 60-Hz line. The secondary is protected with a single fuse. The common side is grounded. Secondary is 120 volts.

   c. Use a green push-to-test pilot light to indicate control voltage is available when a control disconnect switch is closed.

   d. Use a two-position selector switch to select either of two solenoid circuits.

   e. With the two-position selector switch set to position 1, operate a CYCLE START push-button switch.

      1. After a time delay, solenoid A is energized. Operating a REVERSE push-button switch will deenergize solenoid A.

   f. With the two-position selector switch set to position 2, the same CYCLE START push-button switch is operated. Solenoid B is energized immediately.

      1. After a time delay, solenoid C is energized. Operating the same REVERSE push-button switch deenergizes both solenoids B and C.

7. Use graph paper to draw a curve showing the relationship between plunger position and solenoid coil current as the plunger moves from open gap (full out) to closed gap (full in).

8. Why should the duty cycle of a solenoid be known when applying it on a specific application? What precautions should you take?

9. In Figure 5–2, the START push-button switch is operated. Time-delay relay coil 1TR is energized.

   1. Time-delay relay contact 1TR-1 is N.O. on-time delay.

   2. Time-delay relay contact 1TR-2 closes, energizing relay 1CR.

   3. Relay contact 1CR-1 closes, energizing solenoid A.

If the START push-button is released immediately after operating, will the circuit always remain energized? Explain.

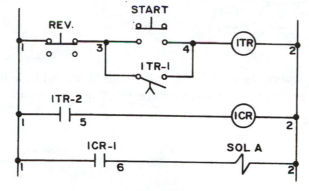

Fig. 5–2

10. Using a step-by-step format (as used in the text), explain the operation of the circuit shown in Figure 5–3.

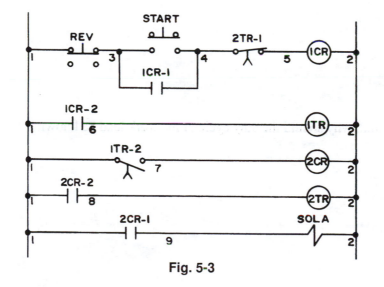

Fig. 5-3

11. What is the difference between "sealed or holding current" and inrush current in an AC solenoid?

12. What type of valve is generally used for open-loop speed control of actuators? What type of valve is mainly used for high-performance closed-loop control of speed or position?

13. If you need a solenoid valve where the spool has to be positioned on either side of center, can you use a single proportional solenoid valve?

14. In using a solenoid, why should the duty cycle of the work load be known?

15. In using a double solenoid valve with both solenoids deenergized, what centers the valve spool?

16. Can a contact on a time-delay relay in a circuit using solenoids be used in an interlock circuit around the START push-button switch?

17. In the circuit shown in Figure 5–4, what action will result if the time-delay relay contact TR-1 is changed to time-delay-after-deenergizing type?

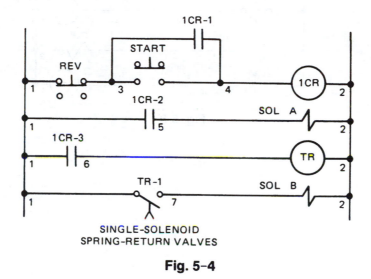

SINGLE-SOLENOID
SPRING-RETURN VALVES

Fig. 5–4

18. In a force motor, what are the results of reversing the direction of current?

19. In the circuit shown in Figure 5–5, add two yellow push-to-test pilot lights that will indicate when solenoids A and B are energized.

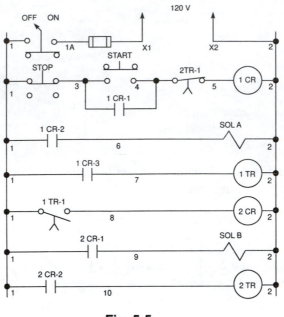

Fig. 5-5

20. Why are laminations used in the iron frame of a solenoid?

21. If an AC solenoid is operating on DC current, what is the difference between the inrush and holding current?

22. What problem might you encounter in operating a DC solenoid on AC current?

23. Under what conditions can a solenoid coil serve as a dual frequency coil?

24. Draw a circuit that will permit fast deenergizing of a solenoid coil for resistance measurement.

25. On a given solenoid coil, hot coil resistance, Rh, divided by cold coil resistance, Rc, is 1.4. Using the chart shown in the textbook Figure 5-6 (below), what is the coil temperature rise?

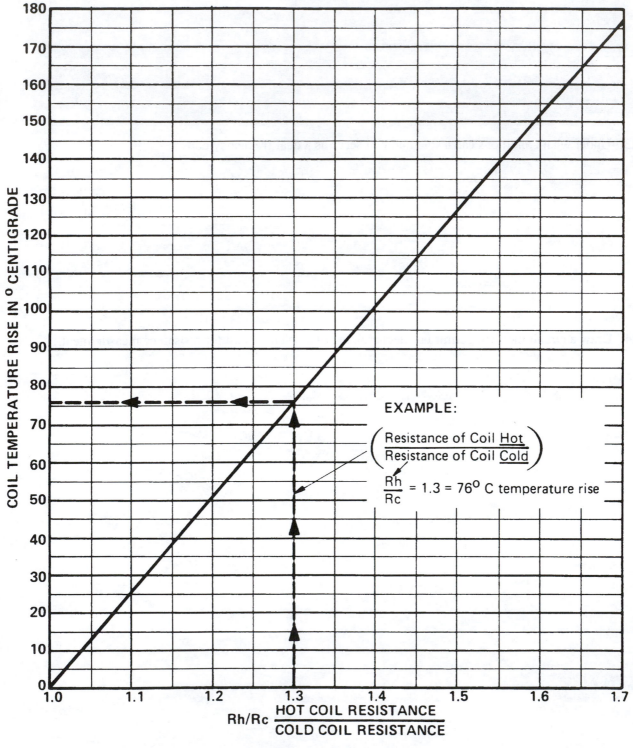

Textbook Fig. 5-6

TRUE AND FALSE STATEMENTS FOR CHAPTER 5

On the line provided, write TRUE if the statement is true or FALSE if the statement is false.

_____ 1. Proportional valves, which use a proportional solenoid to directly or indirectly position the main spool, are used normally for open-loop speed control of actuators.

_____ 2. Force motors give a more linear force/current relationship than proportional solenoids.

_____ 3. In small solenoids, the holding or sealed current is about the same as the inrush current.

_____ 4. Generally two directions of motion are required for proportional solenoids.

_____ 5. In calculating the pull required for a given application for a solenoid, it is generally advisable to underrate the solenoid by at least 10%.

_____ 6. In using oil-immersed-type solenoids, the heat dissipation and wear conditions are improved.

_____ 7. When using servo valves, it is important that the spool is always positioned to two specific positions.

_____ 8. The ratio of inrush current to sealed current in a solenoid generally ranges from 5:1 in small solenoids to 15:1 in large solenoids.

_____ 9. The pull in a solenoid is always constant, regardless of the size.

_____ 10. Force motors give a more linear force/current relationship than proportional solenoids; therefore, there are no limitations on the stroke.

Chapter 6
Types of Control

1. What is the function of an amplifier in closed-loop control?

2. What is meant by the derivative term?

3. In Figure 6–1, what element is missing from the closed loop control diagram?

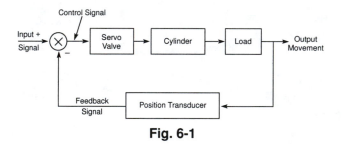

Fig. 6-1

4. In fast-acting systems, what may be the effect of adding the integral term?

5. In open-loop control, what conditions dictate the functions of a machine or process in reaching its preset position or condition?

6. What problems may exist in a machine or process operation that will affect accurate results?

7. What component serves as a major difference between open-loop and closed-loop control?

TRUE AND FALSE STATEMENTS FOR CHAPTER 6

On the line provided, write TRUE if the statement is true or FALSE if the statement is false.

_____ 1. With proportional control, the control signal equals the input signal minus the feedback signal multiplied by the gain.

_____ 2. In many control systems, the load will generally move instantaneously to follow the input signal.

_____ 3. It may be possible to speed up the response of a control system by increasing the gain of the amplifier.

_____ 4. Normally, the proportional signal and derivative signal will have independent gain adjustments to allow the system to be tuned for the best results.

_____ 5. When you have a vertical load situation, any resulting error can generally be compensated for by increasing the gain of the amplifier.

_____ 6. In closed-loop control, where a steady state error exists, the integral term will decrease until the amplifier makes up the difference.

Chapter 7
Motion Control

1. Draw a sketch on graph paper to show the difference between differential travel and total travel of a limit switch.

2. In the circuit shown in Figure 7–1, explain why solenoid A cannot be energized when the START push-button switch is operated if the piston is at position X to start.

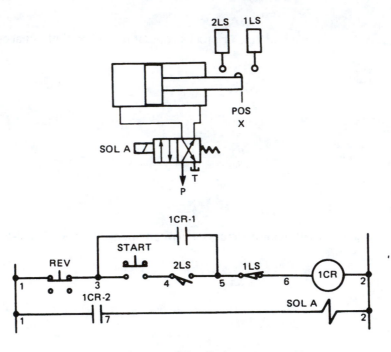

Fig. 7–1

3. Explain how the vane switch operates.

4. In the circuit shown in Figure 7–2, what is the function of the normally closed relay contact 1CR-3?

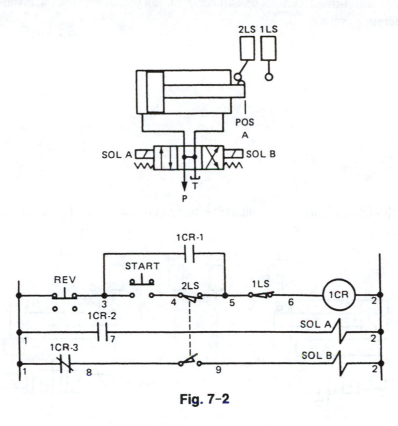

Fig. 7–2

5. In the circuit shown in Figure 7–2, when is solenoid B deenergized?

6. Refer to the piston-cylinder arrangement shown in Figure 7–2. Use graph paper and design a circuit that will require the piston to be in position A for start condition. Solenoid A will be deenergized when the piston reaches and operates limit switch 1LS. After an adjustable time delay, solenoid B is energized. Solenoid B will be deenegized when the piston returns to position A. Provision must be made to return the piston to position A from any point in its forward motion.

7. Explain how the capacitive-type proximity switch operates.

8. Using a linear position-displacement transducer, how can you determine the distance of the external magnet from a reference point?

9. In the circuit shown in Figure 7–3, explain the function of the normally closed contacts on limit switches 2LS and 4LS.

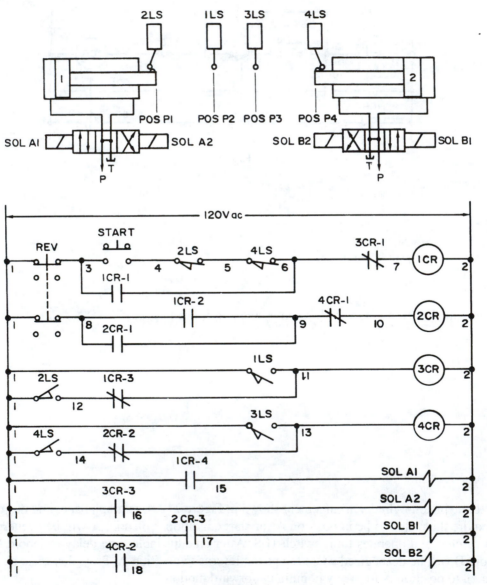

Fig. 7–3

10. In the circuit shown in Figure 7–4, what is the function of the normally closed relay contact 2CR-1?

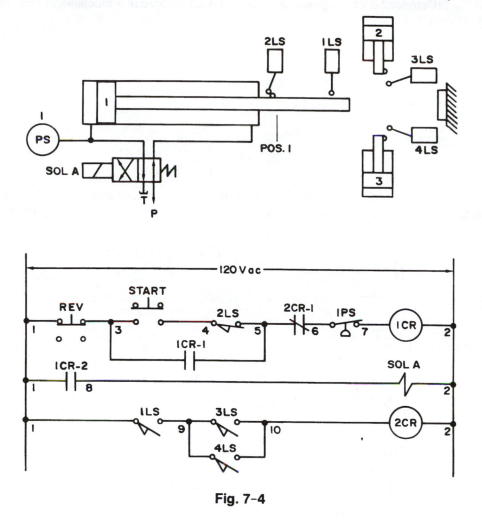

Fig. 7–4

11. Refer to Figure 7–3. Use graph paper and redraw the circuit. Add an additional limit switch 5LS in an operative position between 2LS and 1LS. Now, piston #2 is to move to the left when piston #1 operates limit switch 5LS. What changes would you make in the circuit (Figure 7–3) to accomplish this new operation?

12. The output of a linear position-displacement transducer represents

a. incremental indication, or b. absolute position?

13. Explain the difference between operating force and release force in a mechanical limit switch.

14. In the circuit diagram shown in Figure 7–5, assume the cycle has progressed past 3LS and solenoid B is energized. Will the operation of the REV push-button switch reverse piston B? Explain.

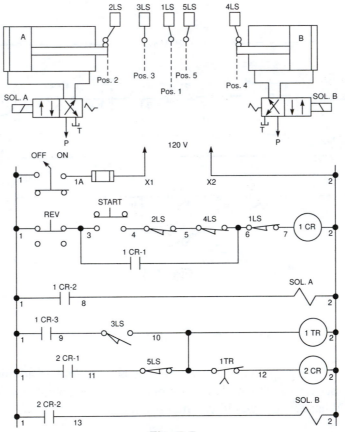

Fig. 7-5

15. Explain Hall voltage.

16. Explain how the capacitive limit switch operates.

17. What can you say about the accuracy and response time for a vane limit switch?

18. What are some of the problems when using a rotary potentiometer as a transducer?

19. Can the transducer shown in Figure 7–6 be used in a closed-loop control circuit? _____

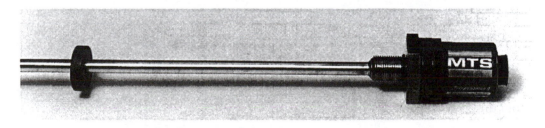

Fig. 7-6 *Courtesy of MTS Sensors Division*

20. Explain the RVDT as shown in Figure 7–7.

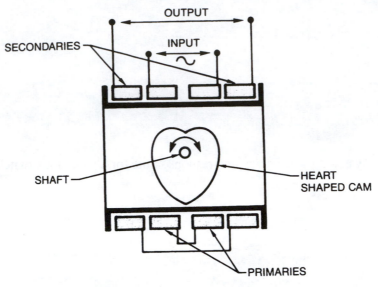

Fig. 7-7 *Courtesy of Vickers, Inc.*

21. What are two types of rotary encoders?

22. List four different types of photoelectric devices.

23. In using a SLO-SYN AC synchronous motor, how can increased holding torque be obtained?

24. What are some of the advantages in the use of a SLO-SYN DC stepping motor as an actuator in digital-controlled positioning?

25. Figure 7–8 illustrates a mass moving horizontally and driven by a rack and pinion or similar device. You are supplied the following information:

Weight	10 lb	Time to reach velocity	0.06 sec
Gear pitch diameter	4 in	Pinion inertia	5 lb-in^2
Gear radius	2 in	Motor rotor inertia	3 lb-in^2
Velocity	12 ft/sec		

Calculate the torque required to accelerate this system. You may ignore frictional forces.

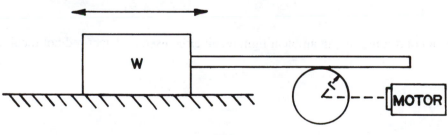

Fig. 7–8

26. In using an inductive sensor, what are some of the variables that determine the distance at which detection will take place?

27. List three applications in which you might use an LED indicator.

28. What controls the response time for a proximity switch?

29. Under what conditions can an ambient light receiver be used to detect red hot metal or glass?

30. What are fiber optics?

31. What are some applications in which air flow monitors are used?

32. Draw a motion profile of a cylinder rod with the following specifications:
   - Extend at acceleration rate of 2" per second per second, until it reaches LS1 when the speed remains constant at 6" per second.
   - When the cylinder rod actuates LS2, it begins to decelerate at 2" per second per second until it stops.
   - When the cylinder is retracted it accelerates at 1" per second per second until it reaches LS3 where it maintains a constant speed of 3" per second.
   - At LS4 it decelerates at 1" per second per second until it stops.

TRUE AND FALSE STATEMENTS FOR CHAPTER 7

On the line provided, write TRUE if the statement is true or FALSE if the statement is false.

_____ 1. The cams on a rotating cam limit switch are independently adjustable for operating at different locations within a 360-degree rotation.

_____ 2. Pretravel on a limit switch is the distance from the free position to the limit position.

_____ 3. The symbol for the normally closed limit switch is the same as the normally open limit switch held closed.

_____ 4. Inductive and capacitive type proximity limit switches use an oscillator in their output circuits.

_____ 5. Rotary potentiometers can be used for measuring angular positions.

_____ 6. The incremental encoder transmits a specific quantity of pulses for each revolution of a device.

_____ 7. Fiber optics are never used for sensing objects in areas of high temperature.

_____ 8. The flow monitor uses a special valve to sense flow of a medium.

_____ 9. The application of angular position displacement transducers is limited to a few degrees of a revolution.

_____ 10. Inductive and capacitive type proximity limit switches have solid-state outputs. They have N.O. and N.C. outputs. In some cases, this is programmable.

Chapter 8
Pressure Control

1. What is the major function of a pressure switch with respect to the electrical control circuit?

2. What tolerance and accuracy would you expect to find in a pressure switch? How are they expressed with respect to the working range of the pressure switch?

3. What care should be taken when using a Bourdon-type pressure switch?

4. What are the contact ratings for snap-action switches generally used in pressures switches?

5. Given a control circuit, Figure 8–1, for the piston-cylinder assembly, you are now to add a second pressure switch (2PS) that is set at a slightly higher pressure than the first pressure switch (1PS). The second pressure switch is to act as a safety device to deenergize solenoid A and energize solenoid B in case 1PS fails to operate at its preset pressure. Use graph paper and design a control circuit to accomplish this safety feature.

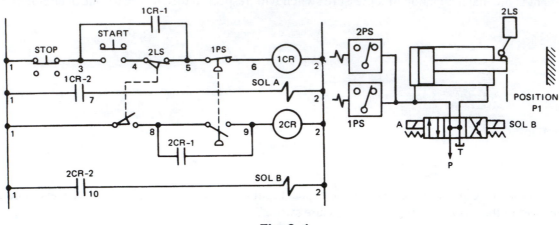

Fig. 8–1

6. In the circuit shown in Figure 8–1, what is the function of relay contact 2CR-1?

7. In the circuit shown in Figure 8–1, add a red push-to-test pilot light that will indicate when the pressure has built to the preset level of 1PS and operated the pressure switch. The light is to be maintained until the piston returns to the original START position. Assume that you have additional unused contacts available on both 1CR and 2CR relays. Use graph paper to draw the circuit.

8. A hydraulic circuit is shown in Figure 8–2 for two piston-cylinder assemblies. Both pistons are to be in a specific position for start conditions. On operating the START push-button switch, piston #1 moves to the right. On engaging a work piece, pressure is built to a preset pressure and then the piston returns to its start position. At the same time that piston #1 builds pressure to a preset level, piston #2 moves to the left. It engages a work piece, builds pressure to a preset level, and then returns to its start position. Use graph paper and design an electrical control circuit that will accomplish the performance described. Note that provisions must be made to return either piston to its start position during their forward travel.

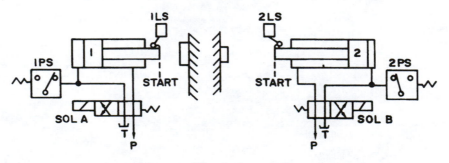

Fig. 8–2

9. In Problem 8, conditions are now changed. Piston #2 is to start its travel to the left at a given position of piston #1 in its forward travel. All other conditions remain the same. What change would you make in the circuit you designed for Problem 8 to accomplish the change in conditions?

10. Differential in a pressure switch is:
   a. the degree of accuracy obtainable.
   b. the yield point less the working range.
   c. the difference between the actuation point and the reactuation point.

11. In a pressure transducer, pressure is applied to a diaphragm through pressure ports and produces:
   a. excessive vibration in the transducer.
   b. minute deflections, which introduces strain to the gauges.
   c. a feedback signal to the diaphragm.

12. In this problem, three piston-cylinder assemblies are used. Three single-solenoid, spring-return solenoid valves are used. See Figure 8–3. Two START push-button switches are used. All pistons must be in their respective start positions for start conditions. On operating both START push-button switches, pistons #1 and #2 advance, clamping a work piece. Both pistons must build and hold pressure to a preset level. When piston #3 reaches its preset pressure level, all pistons return to their respective start positions. Provision must be made to return all pistons to their start positions at any time during their forward travel. Use graph paper and design a control circuit to accomplish these functions.

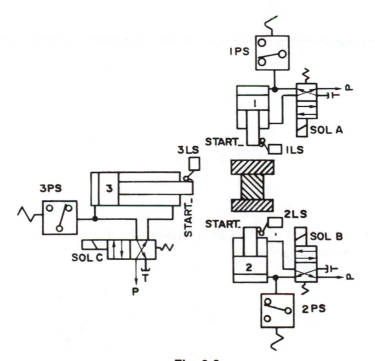

Fig. 8-3

13. What is the approximate tolerance and accuracy found in most pressure switches?

14. If you wanted long life from a sealed piston pressure switch, could 25 pulsations per minute be exceeded?

15. What problem occurs if you exceed the rated pressure when using a Bourdon-tube pressure switch?

16. What elements are used in the solid-state pressure switch for pressure sensing?

17. Use graph paper to draw a four strain gauge Wheatstone bridge circuit indicating the input voltage and the output voltage connections.

18. In the circuit shown in Figure 8–4, what does the broken line that connects the two pressure switch symbols indicate?

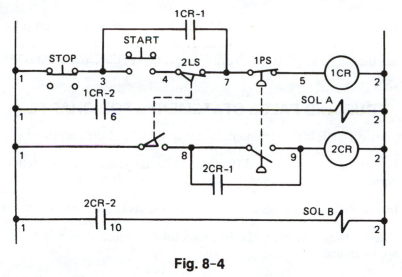

Fig. 8–4

19. In the circuit shown in Figure 8–5, what is the function of relay contact 2CR-1?

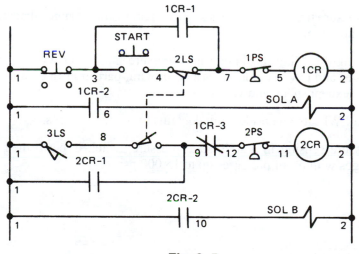

Fig. 8–5

20. In the circuit shown in Figure 8–5, what is the function of N.C. relay contact 1CR-3?

21. Refer to the circuit shown in Figure 8–5. Use graph paper to add a time-delay relay that provides a time delay for the energizing of solenoid B, after solenoid A is deenergized.

22. Using the circuit you designed for question 21, now add a green push-to-test pilot light that will indicate when the preset pressure set on pressure switch 2PS has been reached. The pilot light is to remain energized for a preset time. Use graph paper to draw the portion of the circuit that is involved with the change.

TRUE AND FALSE STATEMENTS FOR CHAPTER 8

On the line provided, write TRUE if the statement is true or FALSE if the statement is false.

_____ 1. In a pressure switch, the differential is the range between the actuation point and the yield point.

_____ 2. In a sealed-piston pressure switch, the seal saves the cost of installing return lines.

_____ 3. Due to its construction, the Bourdon-tube pressure switch is only available for 0–150 psi applications.

_____ 4. In a diaphragm-type pressure switch, pressure is applied against the full area of the diaphragm.

_____ 5. A circuit designer will find it easy to use two independent pressure switches within the same control circuit.

_____ 6. Diaphragm reflection in a pressure transducer results in a digital (millivolt) output that is proportional to the pressure.

_____ 7. A pressure transducer utilizes semiconductor strain gauges that are epoxy-bonded to a metal diaphragm.

_____ 8. The rigid sensing diaphragm in a pressure transducer with bonded semiconductor strain gauges and low mass provides a no-moving-parts transducer that is resistant to most problems except shock and vibration.

_____ 9. A full Wheatstone bridge unit contains four strain gauges.

_____ 10. Pressure sensing in a solid-state pressure switch is performed by semiconductor strain gauges with proof pressures up to 15,000 psi.

Chapter 9
Temperature Control

1. What factors should be considered when selecting a temperature controller?

2. What is meant by operating differential in a temperature controller?

3. Describe the general construction of a RTD unit that is used to sense temperature. Describe how it functions.

4. What basic problems do you find present when using a millivoltmeter-type temperature controller?

5. What are some of the problems that can exist when using a phase-angle fired controller?

6. In the circuit shown in Figure 9-1, add a push-to-test green pilot light that will indicate when the temperature is below 80°F. Assume you have unused N.O. auxiliary contacts available on contactor 10CR. What is the function of the N.C. auxiliary contact on contactor 10CR? Use graph paper to draw the circuit.

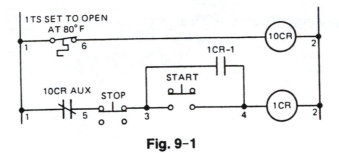

Fig. 9-1

7. In the circuit shown in Figure 9–1, assume that you do not want to energize the coil of relay 1CR until the temperature of the fluid power in the system has reached 70°F. After the temperature has reached 70°F, and relay 1CR is energized, the temperature can drop below 70°F without deenergizing relay 1CR. Assuming that you can install a temperature switch in the operating fluid tank, use the contacts on this switch to accomplish this control function. Use graph paper to draw the circuit.

8. In the circuit shown in Figure 9–2, assume that you want to supply information as to which of the four bearings exceeded the 130°F setting. You have extra relays and pilot lights available to use. Indicate the color of the pilot light lens you are using. Use graph paper and design the new circuit.

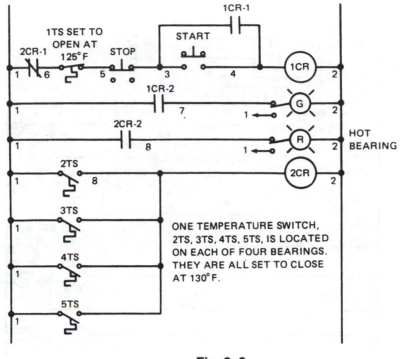

Fig. 9–2

9. A temperature control circuit is shown in Figure 9–3. Contactor 10CR controls the energizing of heating elements located in the operating fluid tank. With the selector switch set to ON and the temperature of operating fluid at or below 75°F, N.C. 1TS thermostat contact is closed, energizing contactor coil 10CR. 10CR contacts close, energizing the heating elements. During the normal operation of the machine, it is found that temperature will fluctuate between 70° and 100°F. The fluid temperature should never exceed 125°F. Explain, by writing a sequence of operations, how you think this circuit will control the fluid temperature in a range of approximately 70° to 100°F.

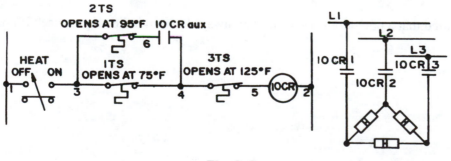

Fig. 9–3

10. Figure 9-4 shows a piston-cylinder assembly, the hydraulic circuit and electrical circuit. Note that a heat control circuit is included. Using a step-by-step format, explain how this circuit operates. Include in your explanation the requirements for starting each cycle and indication that a cycle is completed.

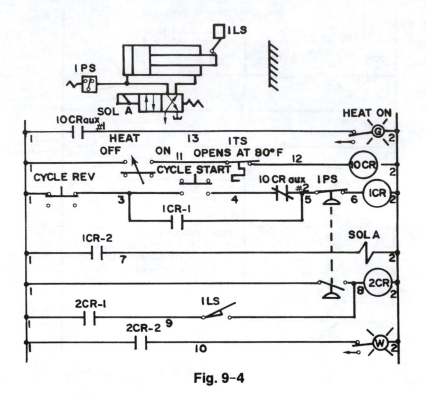

Fig. 9-4

11. Is it required that a potentiometric controller always have an indicating device? If not, explain.

12. You have a single piston-cylinder assembly as shown in Figure 9–5. The piston must always be at a given position for start conditions. On operating the CYCLE START push-button switch, the piston moves to the right until it operates limit switch 1LS. The piston then returns to its start position. The temperature of the operating fluid must be at 70°F or above for starting conditions. However, after starting a cycle, it is not required that the temperature be maintained at 70°F or above during the cycle. If the temperature does drop below 70°F during a cycle, the next cycle cannot be started until the temperature again rises to or above 70°F. If the temperature exceeds 125°F at any time, the cycle will be interrupted (relay and solenoid deenergized) and a new cycle cannot be started until the temperature has dropped below 125°F. Use graph paper and design a control circuit that will satisfy these conditions.

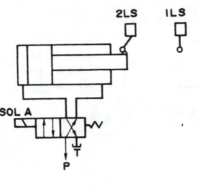

Fig. 9–5

13. What is a thermistor?

14. What are some of the disadvantages in using contactors for switching power?

15. What are some of the advantages in using zero-crossing or zero-crossover fired power controllers?

16. In the schematic drawing for a typical system shown in Figure 9–6, what is the function of the temperature controller?

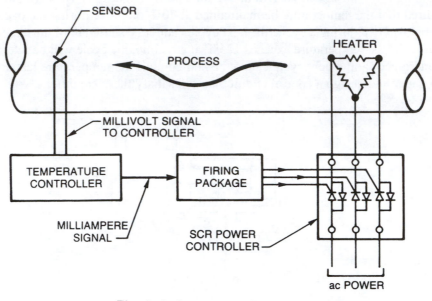

Fig. 9-6 *Courtesy of Chromalox*

17. In Figure 9–7, what is the function of the N.C. auxiliary contact 10CR?

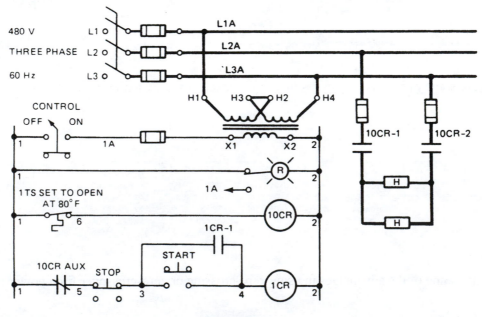

Fig. 9-7

18. In Figure 9–8, what is the function of N.C. relay contact 2CR-1?

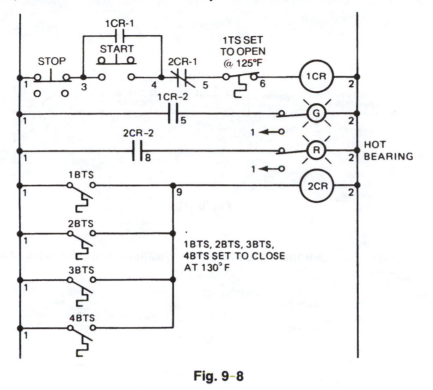

Fig. 9–8

19. Using the circuit shown in Figure 9–9, add an additional pyrometer and heating system, connected in a similar manner as shown in Figure 9–9. However, the second heating system is not to be energized until a preset time after the original system has been energized. Use a time-delay relay.

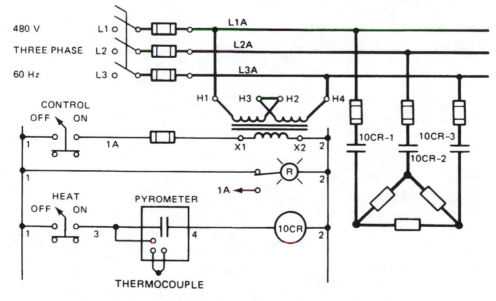

CONDUCTORS CARRYING LOAD CURRENT AT LINE VOLTAGE ARE DENOTED BY HEAVY LINES.

Fig. 9–9

20. In the circuit shown in Figure 9–10, explain the function of the three temperature switches.

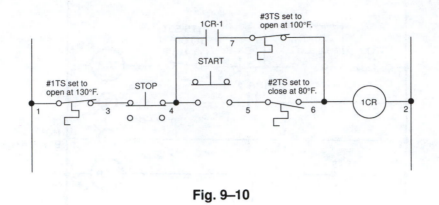

Fig. 9–10

21. In using a potentiometric controller, is temperature indication required to have proper temperature control?

22. Use graph paper to draw two temperature-time diagrams showing the difference between contactor control and SCR control.

TRUE AND FALSE STATEMENTS FOR CHAPTER 9

On the line provided, write TRUE if the statement is true or FALSE if the statement is false.

_____ 1. Overheating in a control cabinet can cause false tripping of protective devices.

_____ 2. The response to a temperature change may vary between one type of sensing element and another.

_____ 3. The combination of copper-iron forms one of the better types of thermocouples.

_____ 4. One of the major sources of trouble with the millivolt-type pyrometer is the adverse effect of rapid changes in temperature.

_____ 5. Mechanical contactors should be used where cycle times are 15 seconds or longer for reasonable service life.

_____ 6. The auto-reset feature in a pyrometer is sometimes referred to as the integral function.

_____ 7. The bimetal temperature switch operates on the principle of even expansion of two different metals when heated.

_____ 8. The liquid-filled type temperature switch that is self-contained has a relatively slow response time.

_____ 9. The RTD type of temperature sensor is the same as the thermistor except it uses a different metal.

_____ 10. Heat can be generated in a circuit carrying power current due to a loose connection.

Chapter 10
Time Control

1. Three contacts on a reset-type timer are used in a control circuit. The contact symbols appear as shown in Figure 10–1.

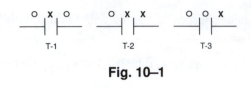

Fig. 10–1

    What information do the circles and X symbols above the timer contact symbols supply to the reader?

2. What controls the point at which contacts open and close on a multiple-interval timer?

3. List some of the important design features found on a solid-state timer.

4. In the circuit shown in Figure 10–2, what change can be made to require that the timer clutch is energized when the START push-button switch is operated? The timer motor will be energized when the pressure builds to the setting on 1PS. Use graph paper to draw the circuit.

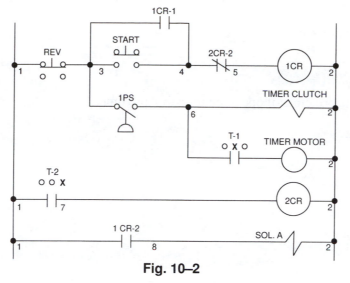

Fig. 10–2

5. In the circuit shown in Figure 10–2, explain the action that would result if the sequence on the timer contact T-2 is changed from OOX to OXO.

6. The circuit shown in Figure 10–2, has been redrawn with a change. Describe the difference in operation of the circuit shown in Figure 10–2 as compared to the circuit shown in Figure 10–3.

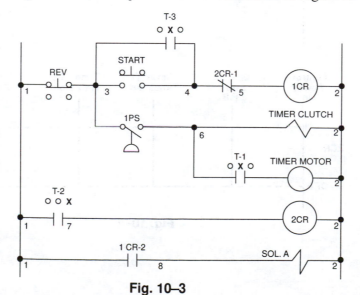

Fig. 10–3

7. In a repeat cycle timer, the cam shaft rotates as long as:

    a. the clutch is energized.

    b. the cams are properly adjusted.

    c. the motor is energized.

8. Explain briefly the principle of operation for a dashpot timer.

9. Given the start of a control circuit as shown in Figure 10–4, complete the circuit, using two reset-type timers to accomplish the following time sequence output to a second relay, 2CR.

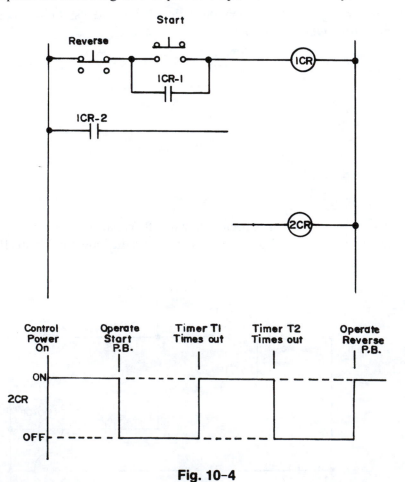

Fig. 10–4

10. What advantage may result in energizing the motor of a reset timer after the clutch has been energized?

11. What is the function of N.C. 2CR-1 relay contact in the circuit shown in Figure 10–5?

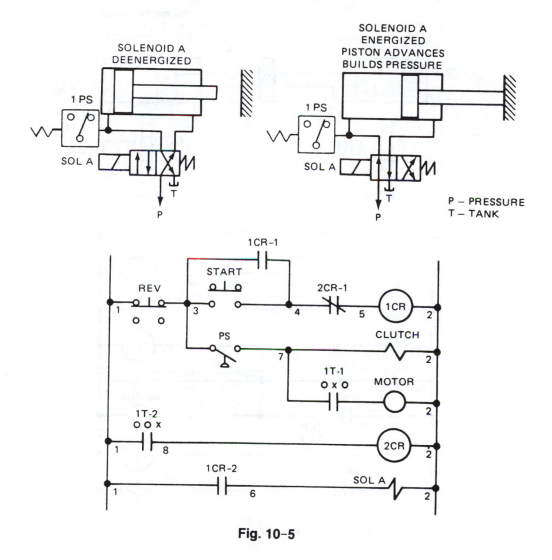

Fig. 10–5

12. In Figure 10–6, what is the significance of the broken line connecting the two pressure switch contact symbols?

13. In Figure 10–6, what is the function of the 2CR-1 relay contact?

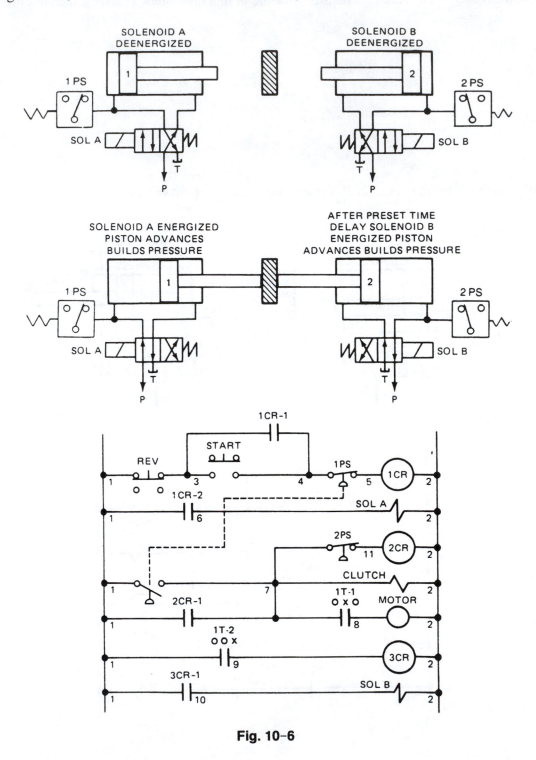

Fig. 10–6

14. In the circuit shown in Figure 10–6, you do not want the circuit to be energized until the operating fluid has reached a temperature of 80°F. After starting, you are not concerned about the temperature falling below 80°F. Where would you place a temperature switch contact in the circuit to accomplish this?

15. In the circuit shown in Figure 10–7, what circuit operation will result if you change the operating sequence of 1T-2 from OOX to OXO?

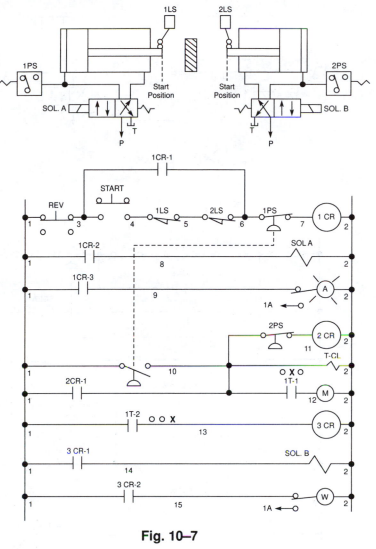

Fig. 10–7

16. In the circuit shown in Figure 10-7, what control operation will result if you change the operating sequence of 1T-1 from OXO to OOX?

17. What is the function of the limit switch contacts on 1LS and 2LS in Figure 10–7?

18. In Figure 10–8, what is the function of N.O. relay contact 4CR-1?

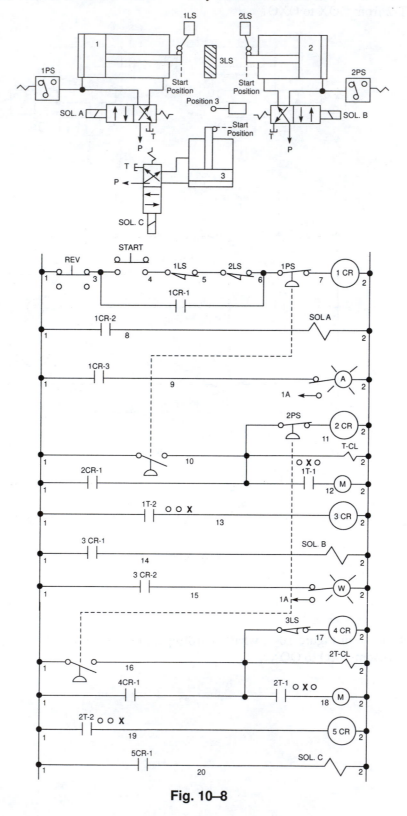

Fig. 10–8

19. In Figure 10–8, will a N.O. held closed 3LS contact provide the same circuit function as a N.C. contact?

TRUE AND FALSE STATEMENTS FOR CHAPTER 10

On the line provided, write TRUE if the statement is true or FALSE if the statement is false.

_____ 1. The multiple interval timer is used extensively in programming control.

_____ 2. The repeat-cycle timer is used to control a number of electrical circuits in a predetermined sequence.

_____ 3. In a reset timer, the clutch and motor are always energized at the same time.

_____ 4. In some solid-state timers, an inhibit function is available.

_____ 5. In the circuit shown in Figure 10–6, you will use a N.O. 2PS pressure switch contact that will close when pressure builds on #2 piston.

_____ 6. In Figure 10–6, when the #2 cylinder extends, it indicates that solenoid B is energized.

_____ 7. In Figure 10–8, the operation of limit switch 3LS will deenergize 2T timer, motor, and clutch.

_____ 8. In Figure 10–8, the changing of 2T-1 timer contact sequence to XOO will have no effect on the circuit operation.

_____ 9. In Figure 10–8, it is not required that the relay contact 1CR-1 in the interlock circuit be connected around both 1LS and 2LS.

_____ 10. The timer and the time-delay relay are the same except that the time-delay relay has a motor but no clutch.

Chapter 11
Count Control

1. When the clutch of a counter is energized, what is the condition of the counter contacts?

2. Approximately how much off time is required between input pulses on a counter to reset?

3. In the circuit shown in Figure 11–1, what is the purpose of the limit switch 2LS normally closed contact?

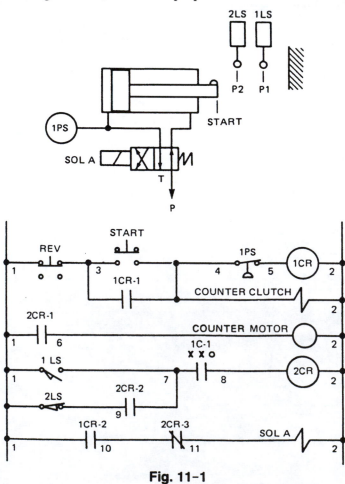

Fig. 11–1

4. In the circuit shown in Figure 11–1, what is the purpose of the normally closed relay contact 2CR-3?

5. In the circuit shown in Figure 11–1, replace the pressure switch with a limit switch that operates at some point to the right of limit switch 1LS. Also provide visual means of indicating when the counter has counted out. Use graph paper to redraw the circuit.

6. Is the typical electromechanical counter generally available with analog set and readout dials?

7. In the circuit shown in Figure 11–2, what is the function of the limit switch contact 2LS?

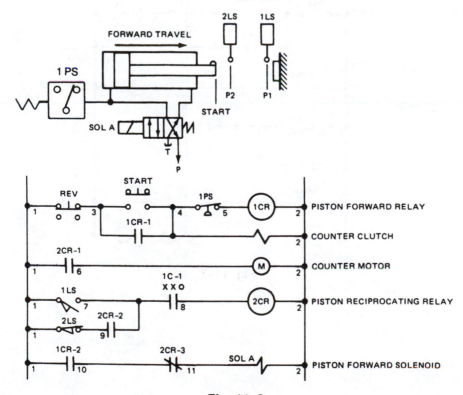

Fig. 11-2

8. What change would you make in the circuit as shown in Figure 11–2 to make the final reversal from position rather than from pressure?

9. In the circuit shown in Figure 11–2, could you replace relay contact 2CR-1 with a 1CR-2 contact and receive the same operating sequence? Explain.

10. Use graph paper and add a push-to-test pilot light to Figure 11–3 to indicate when the pressure has built to a preset level.

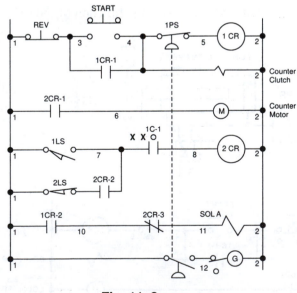

Fig. 11–3

11. Using the circuit you designed for question 8, you now want to energize a second piston-cylinder circuit. The #2 piston is to start forward when the 3LS limit switch is operated. It later reverses to its initial position after operating a limit switch 4LS. Use graph paper to draw the circuit.

12. In the circuit you designed for question 11, you now want to delay the starting of the #2 piston until after a preset time. Use graph paper to design a circuit filling this requirement.

TRUE AND FALSE STATEMENTS FOR CHAPTER 11

On the line provided, write TRUE if the statement is true or FALSE if the statement is false.

_____ 1. With the electromechanical counter, the stepping motor advances one step each time the motor is energized.

_____ 2. The operating contact sequences for an electromechanical counter are Reset-Counting-Deenergized.

_____ 3. High-speed pulse operation in a solid-state counter is available with 100% accuracy.

_____ 4. In the circuit shown in Figure 11–2, the relay contact 2CR-2 is not required if you use an N.O. pressure switch in its place.

_____ 5. In the circuit shown in Figure 11–2, operation of the REV push-button switch will deenergize solenoid A.

_____ 6. In the circuit you developed in question 12; a time-delay relay could be used to replace the electromechanical timer and accomplish the same results.

Chapter 12
Control Circuits

1. Use graph paper to draw a sequence bar chart for the control circuit shown in Figure 12–1.

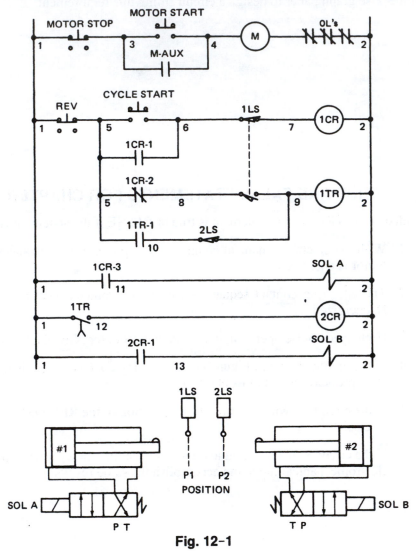

Fig. 12-1

2. What is the function of the normally closed limit switch 2LS in the circuit shown in Figure 12–1?

3. What is meant by an anti-tiedown control circuit?

4. Use graph paper to draw a sequence bar chart for the circuit shown in Figure 12–2.

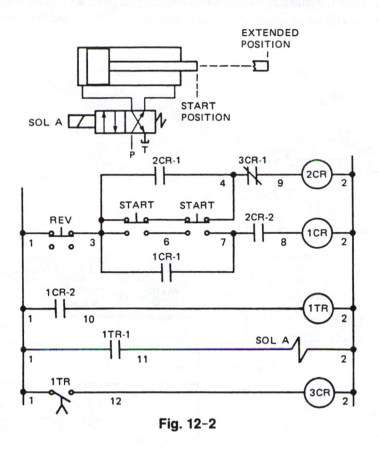

Fig. 12–2

5. In the circuit shown in Figure 12–2, assume that you have accidentally connected the time-delay-after-deenergizing contact in place of the time-delay-after energizing contact. What resultant action would you expect from this change in the circuit?

6. What is the function of the relay contact 2CR-2 in the circuit shown in Figure 12–2?

7. In the circuit shown in Figure 12–3, piston #2 is to return to position #2 after building pressure on block B. It is not to return when the timer times out. Assume you have an additional pressure switch available to use. Change the circuit to accomplish these results. You may remove any unused equipment. Use graph paper to draw the circuit.

8. In the circuit shown in Figure 12–3, explain the use of relay contact 3CR-3.

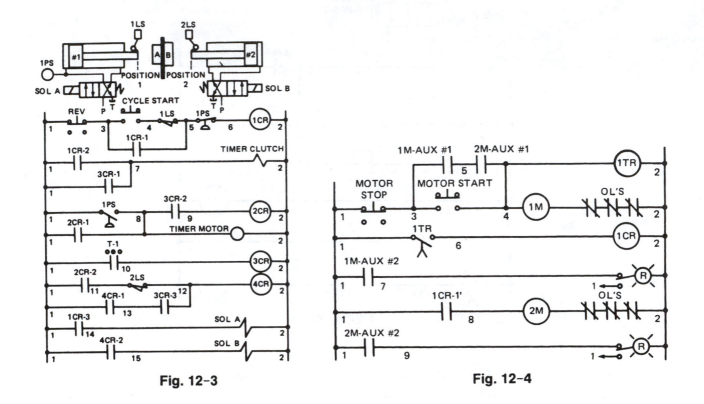

Fig. 12–3   **Fig. 12–4**

9. In the circuit shown in Figure 12–4, add control for a third motor starter. The third motor starter is to be energized after a time delay from the energizing of #2 motor starter (2M). The motor START push-button switch is to be maintained until all three motor starters are energized. The deenergizing of any of the three starters through an overload is to deenergize all three starters. Assume you have extra N.O. interlock contacts available on all three starters. Use only one pilot light to indicate all three starters are energized. Use graph paper to draw the circuit.

10. In the circuit you designed for Question 9, now add the requirement that when the motor STOP push-button switch is operated, motor starters 1M and 2M will be deenergized. However, 3M motor starter is to remain energized for an adjustable time delay. Use graph paper to design the new circuit.

11. Use graph paper to draw a control circuit using REVERSE and CYCLE START push-button switches, a relay, and a solenoid. The solenoid is to be energized when the START push-button switch is operated. The operation of the REVERSE push-button switch is to deenergize the solenoid. On your diagram, divide the input, logic, and output sections with a broken line.

12. Draw a bar graph for the circuit shown in Figure 12-5.

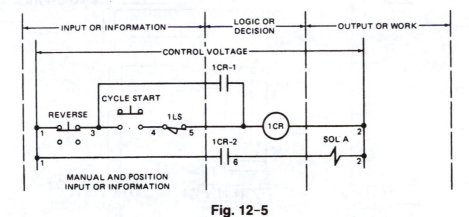

Fig. 12-5

13. In the circuit shown in Figure 12–6, when will the timer motor be energized?

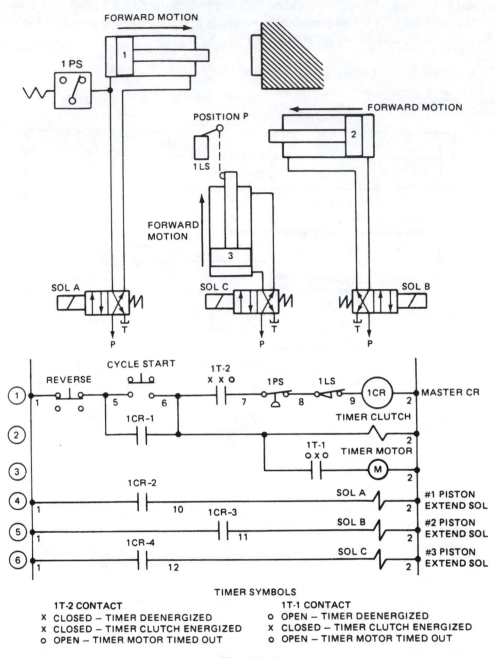

FORWARD MOTION

1 PS

1

POSITION P

1 LS

FORWARD MOTION

2

FORWARD
MOTION

3

SOL A   SOL C   SOL B

T   T   T

P   P   P

CYCLE START

REVERSE   1T-2
               X X o   1PS   1LS
① 1   5   6   7   8   9   1CR   2   MASTER CR

                                            TIMER CLUTCH
       1CR-1
②   2

                         1T-1   TIMER MOTOR
                         o X o
③   M   2

       1CR-2   SOL A   #1 PISTON
④ 1   10   2   EXTEND SOL
                    1CR-3   SOL B   #2 PISTON
⑤ 1   11   2   EXTEND SOL
       1CR-4   SOL C   #3 PISTON
⑥ 1   12   2   EXTEND SOL

TIMER SYMBOLS

1T-2 CONTACT   1T-1 CONTACT
X CLOSED – TIMER DEENERGIZED   o OPEN – TIMER DEENERGIZED
X CLOSED – TIMER CLUTCH ENERGIZED   X CLOSED – TIMER CLUTCH ENERGIZED
o OPEN – TIMER MOTOR TIMED OUT   o OPEN – TIMER MOTOR TIMED OUT

Fig. 12–6

14. In the circuit shown in Figure 12-7, under what condition will the 10CR contactor be energized?

15. In the circuit shown in Figure 12-7, what is the function of the N.C. 3CR-1 relay contact?

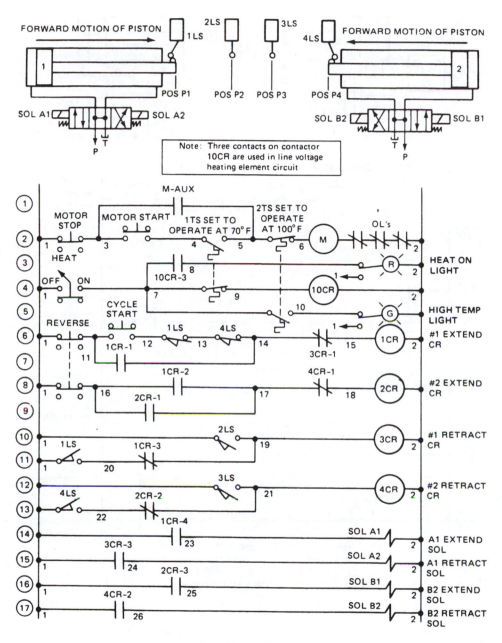

Fig. 12-7

16. In the circuit shown in Figure 12-8, explain how and when #2 piston reverses to its initial start position.

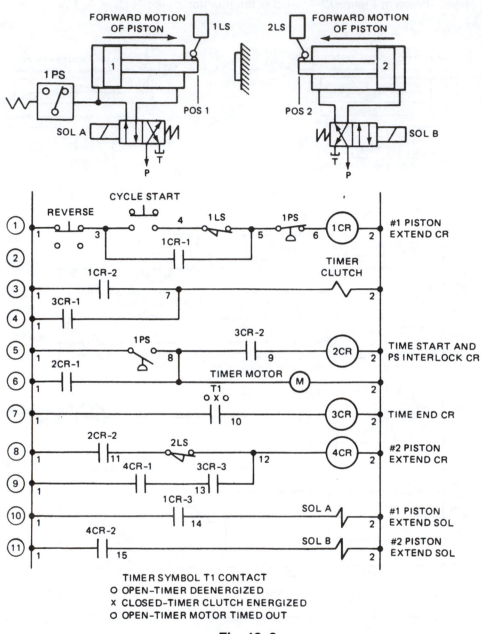

Fig. 12-8

17. In the circuit shown in Figure 12-9, what is the function of the N.O. time-delay contact 1TR?

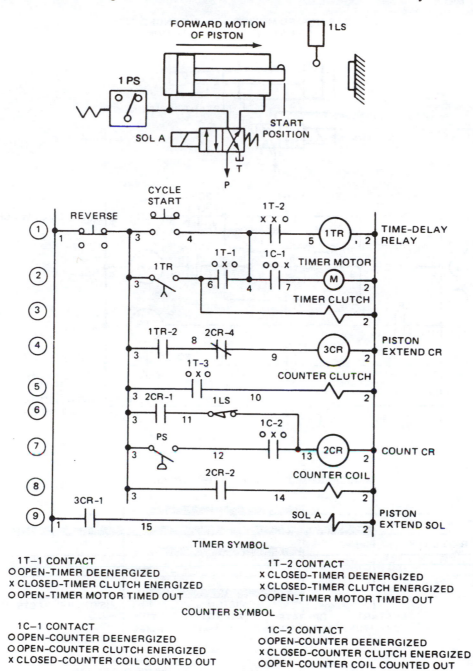

Fig. 12-9

18. In the circuit shown in Figure 12-10, what do you mean by an anti-tiedown circuit? Explain.

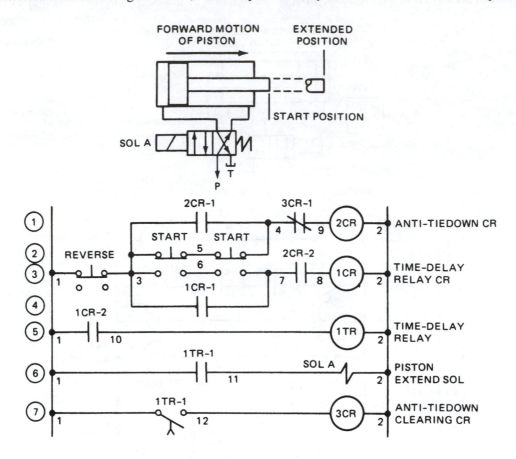

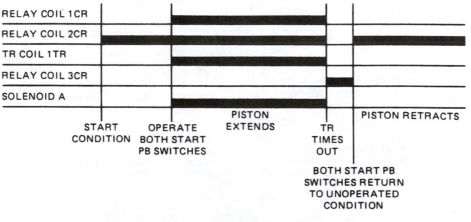

Fig. 12-10

19. In the circuit shown in Figure 12-11, what is the function of the N.O. limit switch contact 5LS?

20. On the circuit shown in Figure 12-11, what is the function of the N.O. held-operated limit switch contact 1LS?

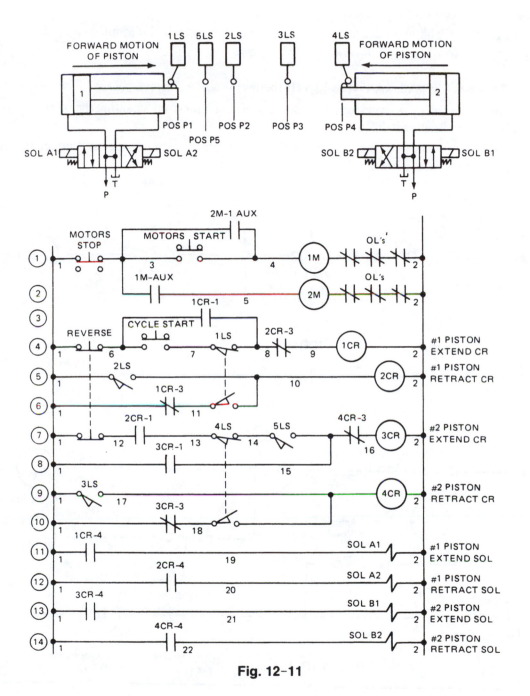

Fig. 12-11

21. An industrial plant has a truck dock at which tractor trailers load and unload materials. The overhead doors (very similar to a garage door found at most homes) require a control system that will control the movement of the door (up or down) as selected by an operator. In addition to the basic control of the door, the controls should allow for:

- automatic reversal of the door if an object is detected in the door's path as it is moving down
- door control pushbuttons are located both inside and outside the building
- use pilot lights to indicate the position of the doors

22. Using the circuit in textbook Figure 12-14B (below)

a. Modify the mixer jog switch circuit to prevent the operator from actuating the mixer motor if the level is below the float switch.

b. Modify the circuit to detect if Solenoid B becomes stuck open to prevent Product A from going directly to the outlet without mixing.

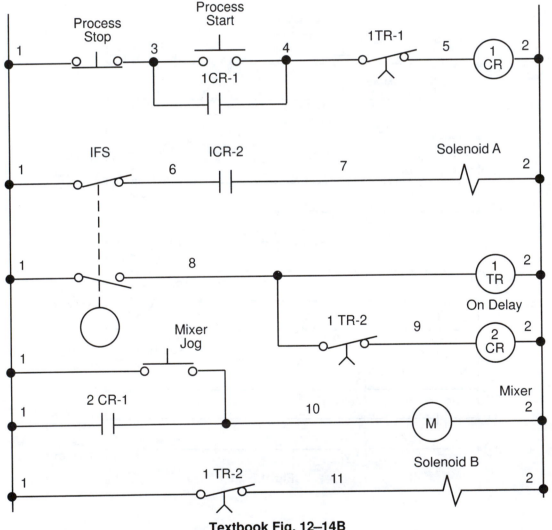

Textbook Fig. 12–14B

23. Using the circuit in textbook Figure 12-15 (below)

   a. Modify the circuit to add an indicator light if the by-pass primary tank solenoid fails.

   b. What is the purpose of N.C. 2CR-3?

   c. What is the purpose of N.O. 2CR-2?

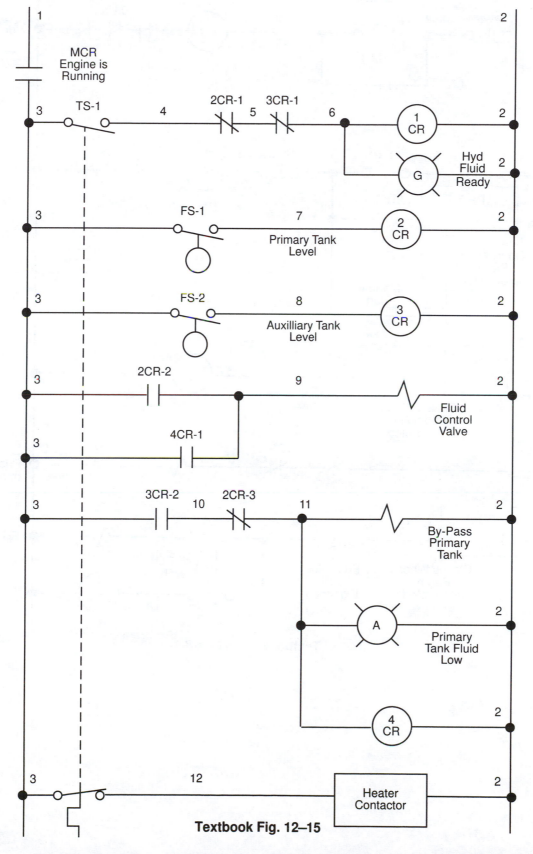

Textbook Fig. 12–15

24. Using the circuit in textbook Figure 12-16B (below)

Modify the circuit to add a glue-dispensing device. Glue is dispensed at the same time the box is being filled. A solenoid valve allows glue to be sprayed on the box lid.

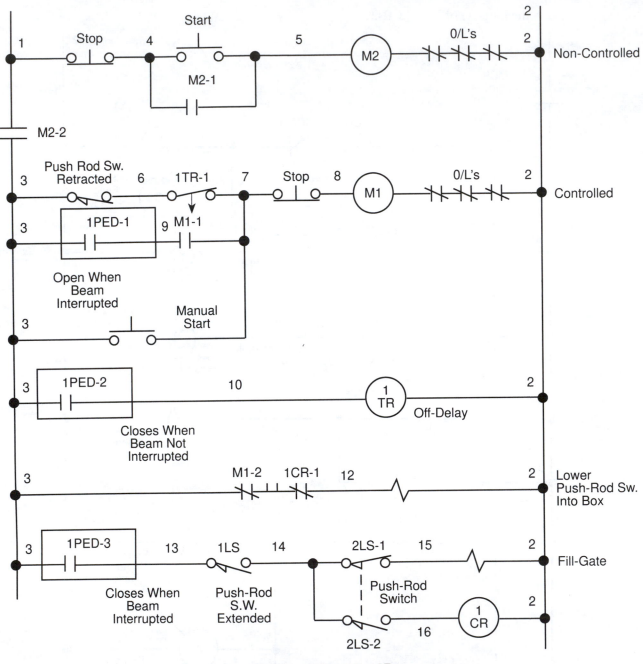

Textbook Fig. 12–16B

25. Modify textbook Figure 12-17D (below)

    a. Add two pilot lights to indicate when the cylinder is at these points:

        PL1   Between P1 and P2

        PL2   Between P2 and P3

    b. Describe how the acceleration rate can be changed.

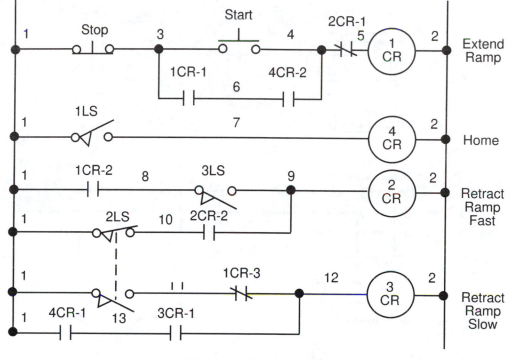

Textbook Fig. 12–17D

TRUE AND FALSE STATEMENTS FOR CHAPTER 12

On the line provided, write TRUE if the statement is true or FALSE if the statement is false.

CONDUCTORS CARRYING LOAD CURRENT AT LINE VOLTAGE ARE DENOTED BY HEAVY LINES.

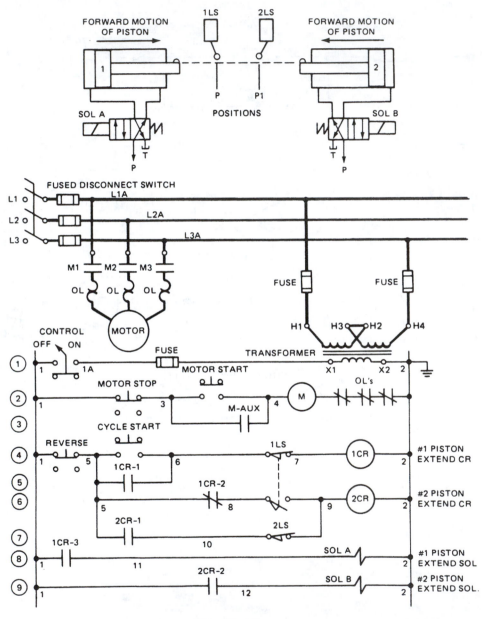

Fig. 12-12

_____ 1. Pyrometer and thermostat outputs provide information for a control circuit.

_____ 2. A bar chart diagram can be used to analyze a control circuit diagram.

_____ 3. In Figure-12–12, the 1CR-1 interlock contact should include the 1LS N.C. contact for proper starting conditions.

_____ 4. In Figure 12–12, the N.C. 2LS contact is not required in series with 1CR-1 for proper operation of the circuit.

_____ 5. In Figure 12–6, the timer motor will be energized when the timer clutch is energized.

_____ 6. In Figure 12–7, relay 1CR will be deenergized when solenoid A2 is energized.

_____ 7. In Figure 12–7, the operation of limit switches 2LS and 3LS will depend on their relative positions and relative velocity of the pistons.

_____ 8. In Figure 12–8, it will be necessary for the pressure switch contact 1PS to close before the timer motor can be energized.

_____ 9. In Figure 12–9, the pressure switch contact 1PS will not operate until the counter has counted out.

_____ 10. In Figure 12–10, if you substitute a N.C. 1TR-2 timing contact for N.C. 3CR-1 relay contact, you can defeat the anti-tiedown feature of this circuit.

Chapter 13
Motors

1. Why are stator laminations insulated before they are "packaged"?

2. What induces a voltage in the squirrel cage rotor?

3. What produces torque in an AC induction motor?

4. What is "slip" and is it important in the operation of an induction motor?

5. Explain the efficiency and starting torque of the resistance split-phase motor.

6. With a capacitor start motor, the addition of a capacitor results in a time phase shift that is much closer to 90° than with the split-phase motor. What can you say about the resulting starting torque?

7. What are some of the real strengths of the permanent split capacitor design?

8. Does the shaded pole motor use a start winding? Explain.

9. What is the synchronous speed of a four-pole, three-phase 60-hertz squirrel cage induction motor? Show your work.

10. Explain the efficiency and starting torque for a shaded pole motor.

11. What is the purpose of a small number of series turns in a stabilized shunt exciting field?

12. What are three commonly used methods of controlling the speed of a DC motor?

13. At constant current, what happens to the horsepower of a DC motor as the armature voltage is increased?

14. What torque in lb.ft. is developed by a 10 HP DC motor running at 1000 rpm?

15. A DC motor runs at 1500 rpm no load and 1250 rpm full load. What is the speed regulation?

16. Why do the main drives of metal-working machines require approximately constant HP?

17. In a brushless DC motor, even though the current required by the motor to develop the torque may be large, why is the actual power used small at low speeds?

18. The brushless DC motor control has an "electronic commutator" fed by an integral encoder mounted on the motor. What is the purpose of this encoder?

TRUE AND FALSE STATEMENTS FOR CHAPTER 13

On the line provided, write TRUE if the statement is true or FALSE if the statement is false.

_____ 1. In a squirrel cage induction motor, the use of solid iron poles will result in overheating and poor efficiency.

_____ 2. In a squirrel cage induction motor, the rotor laminations are punched out of the stator laminations.

_____ 3. In a squirrel cage induction motor, current flowing through the coils of wire embedded in the stator slots creates definite north-south poles.

_____ 4. In a squirrel cage induction motor, the running speed is generally equal to or greater than the synchronous speed.

_____ 5. If you are able to spin a single-phase motor by hand, it will develop maximum torque.

_____ 6. The start winding in a resistance split-phase motor generally has more turns of larger wire than that of the main winding coils.

_____ 7. The capacitor start motor utilizes the same winding arrangement as the split-phase motor but adds a short time-rated capacitor in series with the start winding.

_____ 8. In the permanent split-capacitor motor, continuously rated capacitors are normally provided in small-microfarad, high-voltage ratings.

_____ 9. The shaded-pole motor is the most efficient of all the single-phase motors.

_____ 10. The shaded-pole motor is the most simply constructed and hence the least expensive of the single-phase designs.

Chapter 14
Motor Starters

1. What is meant by a combination motor starter?

2. Why is it that more than one set of overload relays are used in a multispeed motor starter?

3. List four types of reduced-voltage motor starters.

4. What is the function of a timer in the primary resistor-type reduced-voltage motor starter?

5. What consideration must be made in selecting a motor to use with a wye/delta reduced-voltage motor starter? What is the normal starting torque for a motor using a wye/delta motor starter?

6. Use graph paper to draw a control circuit using two full-voltage motor starters. The first motor starter is to be energized when the motor START push-button switch is operated. The second motor starter is to be energized following a short time delay. Provision must be made to deenergize both motor starters at any time. Assume you have a time-delay relay available to use. If either motor starter coil is deenergized due to an overload, it is not required that both motor starter coils be deenergized.

7. Figure 14–1 shows three motor starters connected to start in sequence from a single motor START push-button switch. Explain why you think a 3M auxiliary contact is used to interlock the motor START push-button switch. What action follows the deenergizing of #1M motor start coil due to an overload on #1 motor?

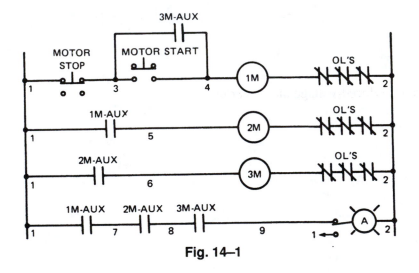

Fig. 14–1

8. In the control circuit shown in Figure 14–1, add an amber push-to-test pilot light that will indicate when all three motor starters are energized. Assume you have unused N.O. auxiliary contacts available on all the starters. Use graph paper to draw the circuit.

9. What is the approximate difference in starting current between a full-voltage starter and a part winding reduced voltage starter in percent of full-load current?

10. Is there any adjustment for the starting current in a solid-state reduced-voltage starter? Approximately what range of starting current in percent of full-load current is covered?

11. Use graph paper to draw an electrical control circuit diagram for a full-voltage reversing motor starter. Explain the function of the electrical interlocks. You do not have to draw the power section of the circuit.

12. Assume you have two motors to start. You are using full-voltage starters. After operating a master motor START push-button switch, there is to be a time delay before the first motor starter coil is energized. After the first motor starter coil is energized, there is to be a second time delay before the second motor starter coil is energized. The master motor START push-button switch is to be held operated until the second motor starter coil is energized. Both motor starter coils can be deenergized at any time by operating a motor STOP push-button switch. Use graph paper and design an electrical control circuit that will accomplish this operation. In your design, if the second motor starter coil is deenergized due to an overload, will both motor starter coils be deenergized? Explain. You do not have to show the power section of the circuit.

13. You are using two motors in a process, each with their own START-STOP push-button switches. If either motor is overloaded, both motors should be deenergized. Use graph paper and design a circuit that will accomplish this result.

14. In the circuit shown in Figure 14–2, how does the control change in respect to the circuit shown in Figure 14–3?

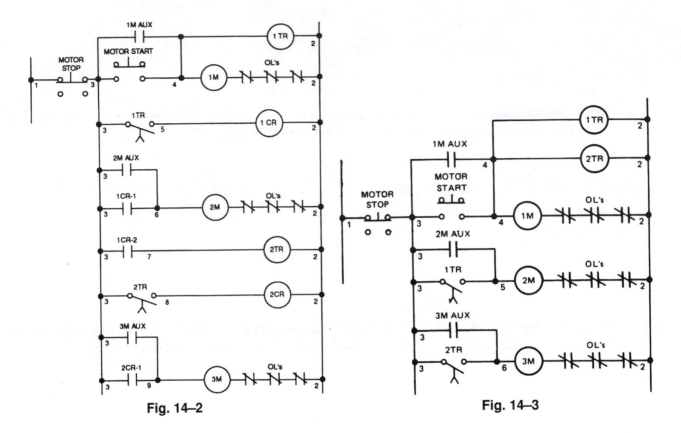

Fig. 14–2   Fig. 14–3

15. What is "locked rotor" current?

16. What is one problem involved with an auto-transformer reduced-voltage starter using open transition?

17. What might be one problem in using a wye-delta motor starter with a standard three-phase squirrel-cage induction motor?

18. In the circuit shown in Figure 14–4, what is the function of the TR timing contact?

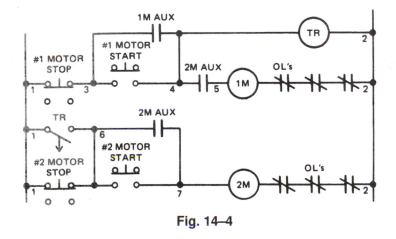

Fig. 14–4

19. In the circuit shown in Figure 14–5, when will #2 motor starter be energized?

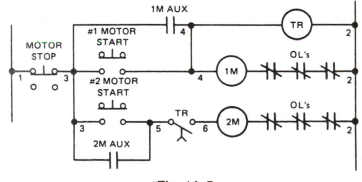

Fig. 14–5

20. Use graph paper to draw a circuit showing how no-voltage or low-voltage protection can be accomplished.

21. In the circuit shown in Figure 14–6, explain how jogging is accomplished.

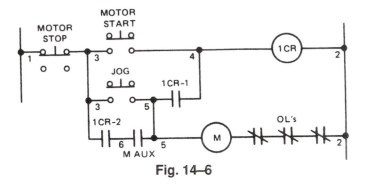

Fig. 14–6

22. Draw a circuit for a full-voltage reversing motor starter. Show both power and control connections.

Name _____

23. In the circuit shown in Figure 14–7, what is the function of N.C. contacts SM and FM?

THE CIRCUIT IS NOT NECESSARILY THAT OF THE STARTER SHOWN

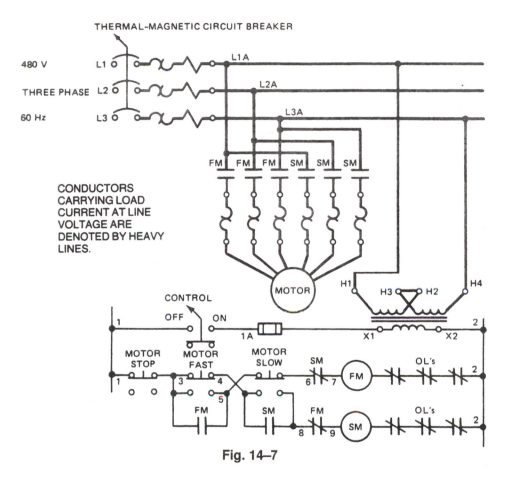

Fig. 14–7

24. Sam has a machine shop with one 30 HP, 480 volt, 3 PH, 60 Hz AC induction motor and one 50 HP, 480 volt 3 PH, 60 Hz AC induction motor. The main power bus in the shop is 480 volts, 3 PH, 60 Hz. He has a ganged 3-pole thermal magnetic circuit breaker to protect his shop power. Each motor is protected by a 3-pole, ganged fused disconnect switch. He wants to control the motor starters at 120 volts, single phase, through a control transformer. The transformer is protected on both the primary and secondary sides with fuses. In addition to the standard start-up motor starter control, he wants an emergency motor stop pushbutton switch with mushroom head to deenergize both motors when it is operated. Draw the power and control circuit for this application.

25. Referring to the conditions outlined in question 24, Sam now finds that he needs an additional 10 HP, 480 volt, 3 PH, 60 Hz AC induction motor to run a conveyor. The motor is to be protected by a ganged 3-pole fused disconnect switch. This motor must be started before the two larger motors can be started. He would also like to have indicating lights available to tell him when the motors are energized. Assume that you have extra interlock contacts available on all the motor starters. All motors will be deenergized when the emergency stop pushbutton is operated. Make the necessary changes to your answer in question 24 to accommodate these new requirements.

26. If the 10 hp motor in question 25 is deenergized due to an overload, how are the other two motors affected?

TRUE AND FALSE STATEMENTS FOR CHAPTER 14

On the line provided, write TRUE if the statement is true or FALSE if the statement is false.

_____ 1. The contact in a thermally responsive overload relay is connected in parallel with the starter coil.

_____ 2. Overload relays can be used for both overload and short-circuit conditions.

_____ 3. In a reversing motor starter, both mechanical and electrical interlocks are used to prevent both starters from closing their contacts at the same time.

_____ 4. If you use the pole-changing method in a two-speed motor, a four-pole motor will run at 1200 rpm.

_____ 5. In using the jogging function in a motor starter, a relay is used to prevent the starter from "locking in."

_____ 6. In Figure 14–8, the N.C. limit switch contact is used as an overstroke limit switch.

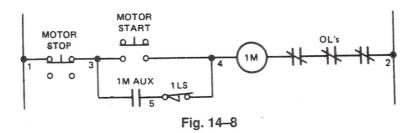

Fig. 14–8

_____ 7. When starting a motor using the reduced-voltage method, you will find that the torque is reduced directly as the voltage is reduced.

_____ 8. Using the wye/delta reduced-voltage starter, torque will be reduced to 50% of normal starting torque on starting.

_____ 9. The area of adjustment for a solid-state reduced-voltage motor starter is greater than any of the other types of reduced-voltage starters.

_____ 10. The use of a master motor stop push-button switch is often used as a convenience or emergency safety. It is generally supplied with a mushroom-head push-button operator.

Chapter 15
Introduction to Programmable Control

1. What is the purpose of indicating lights that are generally found on interface units?

2. What are several basic functions central processing units (CPUs) perform in a programmable controller?

3. Use graph paper to draw a box-form diagram showing the relationship among the following hardware devices: input and output interfaces; memory; processor; power supply; programming equipment.

4. Is the presentation of addresses in programmable controllers the same for all manufacturers? Explain.

5. In what way does the program format vary from one manufacturer to another?

6. What advantages does a printer offer to the user of programmable control?

7. What are some of the applications for programmable control in industry?

8. In the circuit shown in Figure 15–1, what information would you supply through input interfaces and output interfaces if you were to put this circuit on a programmable controller?

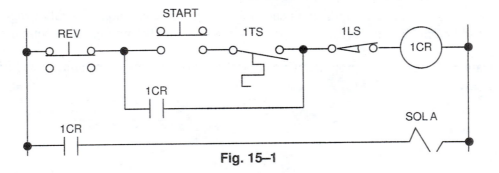

Fig. 15–1

9. Name the primary advantages of using non-volatile Read Write (R/W) memory?

10. What is meant by the term "scanning"?

11. What is an analog signal? Is it ever necessary to convert a digital signal to an analog signal? Explain.

12. What problem can you encounter when using a volatile memory system?

13. What information should be included when documenting the operation of a machine for a programmable controller?

14. Show in block form the three basic divisions of control for the PLC.

15. As used in a PLC output circuit, show a bridge rectifier with resistors to drop the voltage to a required level.

16. What type of battery is generally preferred when a back-up power supply is required? Explain why?

17. What is the major difference between the PROM memory and the CMOS memory?

18. What is the function of a floppy disk drive?

19. When a modem accepts digital data from either a processor or programming terminal, what action does it take?

20. In using modems, what transmission frequencies are generally used over commercial telephone lines?

21. In using a printer, what precautions should be taken with the communication link between the programming terminal and the printer?

22. What precaution should be taken in selecting a power supply for the PLC?

23. Given the control circuits developed and explained in previous chapters, using PLC symbols, convert the following circuits to PLC control

    a. assume all normally open input switches

    b. assume that all input devices must remain as originally designed:

    Figure 15–2

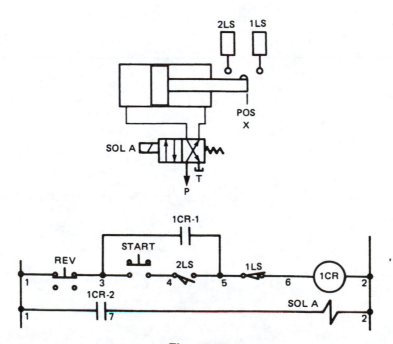

Fig. 15–2

Figure 15–3

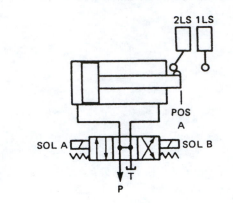

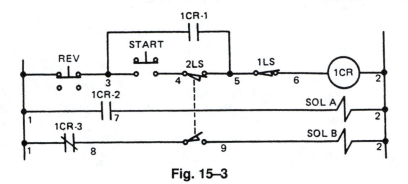

Fig. 15–3

Figure 15–4

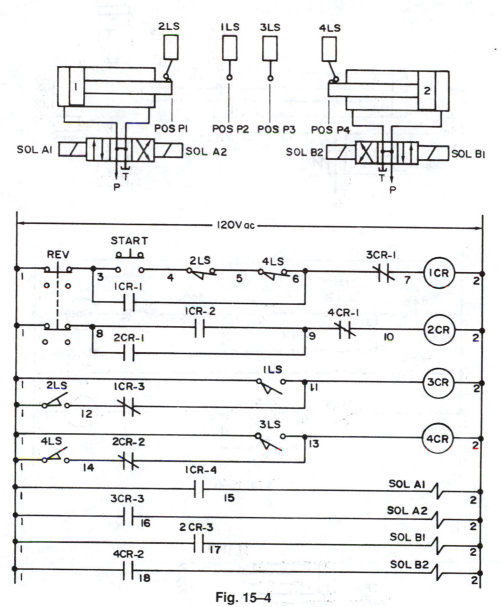

Fig. 15–4

Figure 15–5

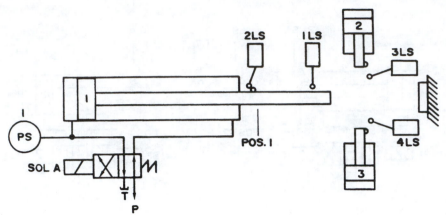

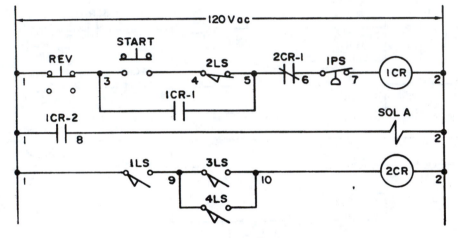

Fig. 15–5

Figure 15–6

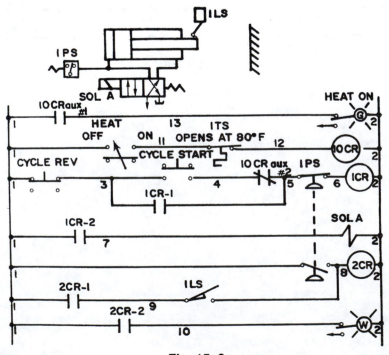

Fig. 15–6

Figure 15–7

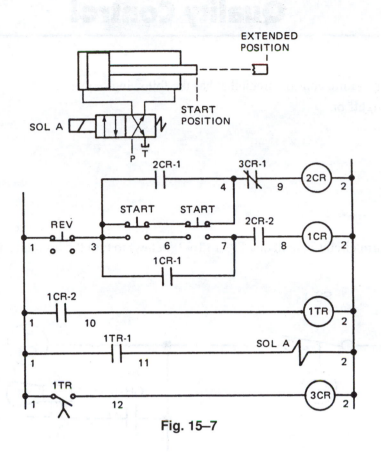

Fig. 15–7

TRUE AND FALSE STATEMENTS FOR CHAPTER 15

On the line provided, write TRUE if the statement is true or FALSE if the statement is false.

_____ 1. With the PLC you will find a more complete division of control than in the electromechanical control.

_____ 2. Process switches such as a pressure switch provide input status to the programmable logic circuit.

_____ 3. Process-indicating devices, such as a pilot light, can be directly driven from the outputs of a PLC.

_____ 4. An advantage of a PLC over a hard-wired relay logic circuit is that a program replaces the physical wiring.

_____ 5. In the input circuit for a PLC, a bridge rectifier is used to convert ac to dc and resistors drop the voltage to the required level.

_____ 6. The LED can be used to isolate the incoming signal to a PLC before it enters the processor.

_____ 7. In the PLC, a microprocessor serves to monitor the status (ON/OFF) of the input devices. It also scans and solves logic of the user program.

_____ 8. Volatile memory is made up of two types: MOS and CMOS.

_____ 9. It is not required that a printer be compatible with a processor.

_____ 10. The use of the data highway in a plant is always limited to 1000–2000 feet of cable length.

Chapter 16
Quality Control

1. Develop a temperature control circuit that has the following sequence:

   Red and blue lights on

   250--------------------

   Blue light on

   200--------------------

   Red light on

   100--------------------

2. Modify the alarm circuit in textbook Figure 16-2 (below) to turn the horn on and off at a rate of 3 sec on and 2 sec off.

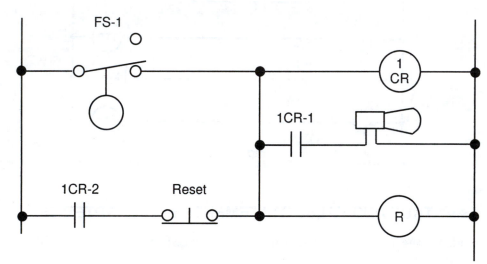

Textbook Fig. 16–2

3. In the application shown in textbook Figure 12-14 (below), add an annunciator panel to provide operator notification of critical events.

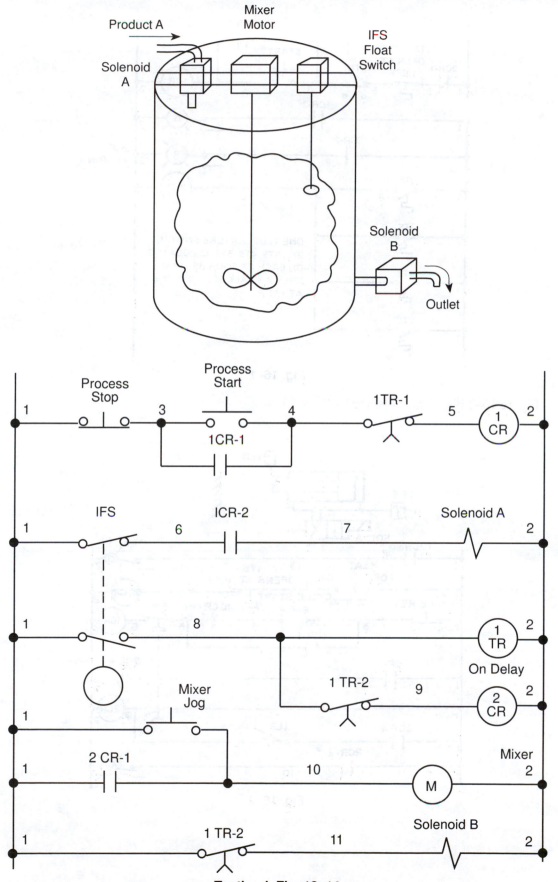

Textbook Fig. 12–14

4. What are the implications of

   a. Figure 16–1 if the thermal switch ITS doesn't close?

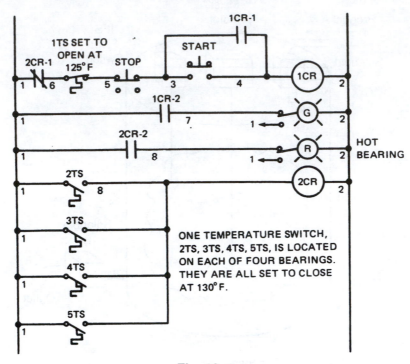

Fig. 16–1

   b. Figure 16–2 if 1TS fails open at 80 F

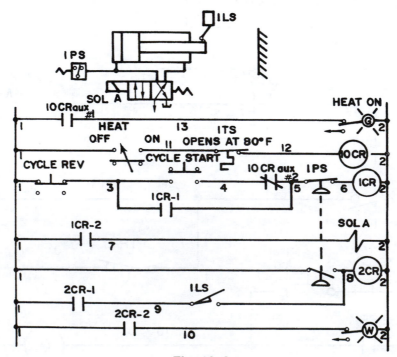

Fig. 16–2

c. Figure 16–3 if 3TS fails

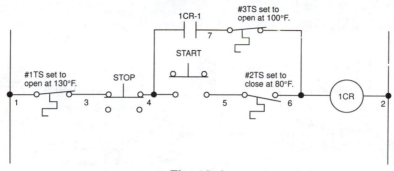

Fig. 16–3

d. Figure 16–4 if PS fails to close

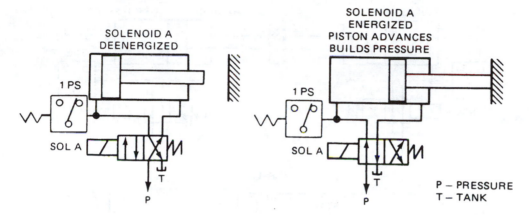

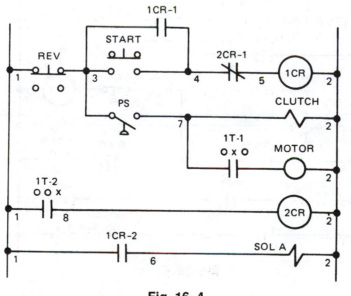

Fig. 16–4

e. Figure 16–5 if 2PS fails to open

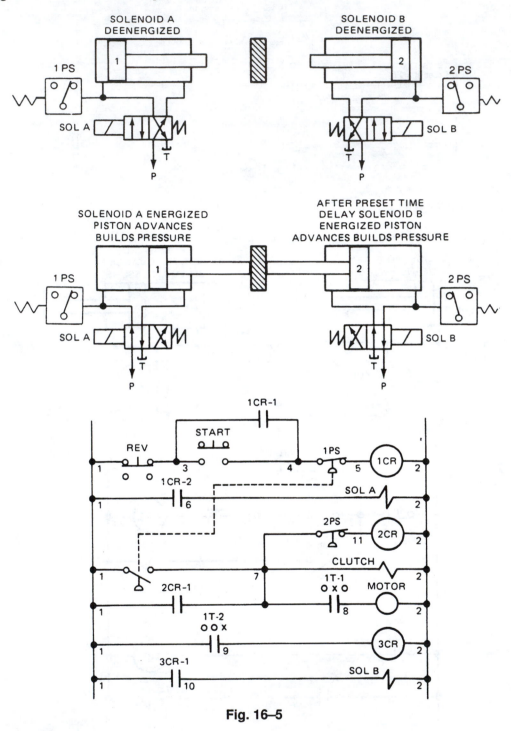

Fig. 16–5

5. Define process tolerance.

6. Give an example of a hand-operated device that can be used to verify dimensions and tolerances.

7. When using a control chart (similar to Figure 6-6 in the textbook), how is the control band determined?

8. Define the term ADC.

9. Provide two examples of problems that can cause a quality control system to become inaccurate.

10. Provide an advantage for using an annunicator panel.

TRUE AND FALSE STATEMENTS FOR CHAPTER 16

On the line provided, write TRUE if the statement is true or FALSE if the statement is false.

_____ 1. Sensors never need calibration.

_____ 2. When tolerance measurements of machined parts begin to exceed standard values, the rejection region of a normal distribution curve will be come larger.

_____ 3. The "within tolerance" region of the normal distribution curve will never be smaller than the rejection region.

_____ 4. In statistical measurements, the value that occurs most frequently is called the mean.

_____ 5. Process measurements larger than the upper control limit could be considered "out of control."

_____ 6. Open-loop control can always provide better control of tolerance than closed-loop control.

_____ 7. A data acquisition system usually can monitor more than one sensor.

_____ 8. Usually it is the machine operator's responsibility to determine tolerance levels for quality.

_____ 9. Only analog instruments can be used to measure process tolerances.

_____ 10. Hand-held instruments are more accurate than digital measurement devices.

Chapter 17
Safety

1. Using textbook Figure 12-16A (below) and assuming the cylinders are most dangerous to an operator when they are extending, add appropriate circuits and alarms to indicate when the cylinders are in motion, both extending and retracting.

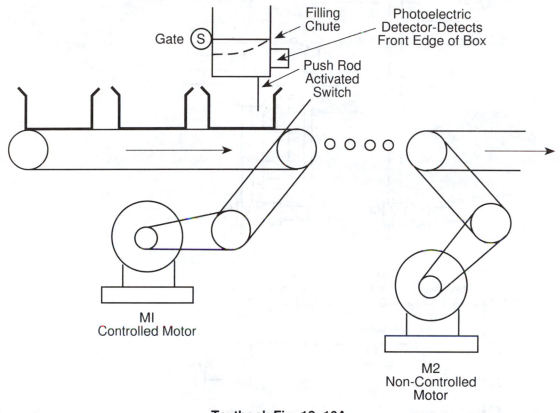

Textbook Fig. 12–16A

2. Using textbook Figure 12-7A (below) and assuming the cylinc
   when they are extending, add appropriate circuits and alarms to in
   both extending and retracting.

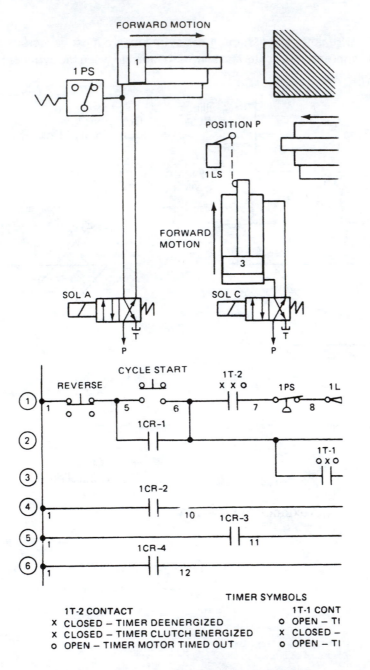

Textbook Fig. 12–7A

3. Refer to textbook Figure 12-8A (below). Modify the circuit so all cylinders will immediately retract if an object enters the volume work area of the cylinders in motion.

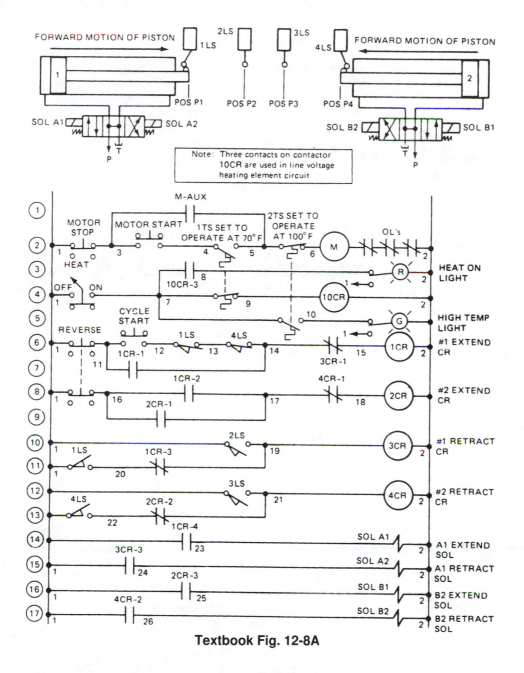

Textbook Fig. 12-8A

4. Refer again to textbook Figure 12-8A. What would occur if:
   a. 3LS fails to close
   b. 5LS fails to close
   c. the NCHO contact of lLS breaks in the open position

5. What is the primary role of a worker safety circuit?

6. What is the purpose of laws relating to safety?

7. Identify three sensors/switches which can be used in safety circuits.

8. List the programmed sequence usually followed in safety circuits.

9. What is the purpose of a diagnostic system?

10. List the benefits of a "safety controller."

TRUE AND FALSE STATEMENTS FOR CHAPTER 17

On the line provided, write TRUE if the statement is true or FALSE if the statement is false.

_____ 1. A significant number of OSHA's general industrial citation point to the lack of appropriate machine guards.

_____ 2. Safety is a process that is constantly being refined and improved.

_____ 3. Electrical power to process machinery may remain on when maintenance workers are working in the area.

_____ 4. Machine safety is always more important than worker safety.

_____ 5. Machines need to be protected from moving elements of other machines.

_____ 6. The first level of operator safety is to stop the motion of any mechanical mechanism where any portion of the worker's body enters into a dangerous area.

_____ 7. Lubricating oil is not a safety factor for machines.

_____ 8. OSHA stands for Occupational Safety and Health Administration.

_____ 9. The volume of space where robotic mechanisms operate are always safe for workers.

_____ 10. Multiple stop devices electrically connected in parallel is a common safety control technique.

Chapter 18
Troubleshooting

1. Describe the problem that develops when there is a loose connection in a power line.

2. What is meant by cross firing in a switch that has double-pole, double-break contacts?

3. Describe a problem that develops in a solenoid, causing it to burn out.

4. Explain one approach in developing a good preventive maintenance program.

5. If you have a control circuit using many components with normally open and normally closed contacts, where is a wiring error most likely to occur? Explain.

6. In Figure 18–1, a circuit is shown with a wiring error. Using a voltmeter, describe how you would proceed to check the circuit for the error.

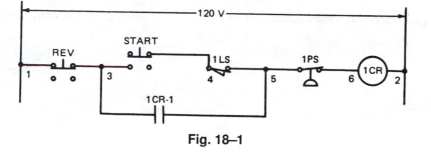

Fig. 18–1

7. List four steps you could take in locating the source of a momentary fault in a control circuit.

8. In the circuit shown in Figure 18-2, describe the problem that can develop after the operation of limit switch 1LS. Can you suggest a change in the circuit that would help? Use graph paper to draw the circuit.

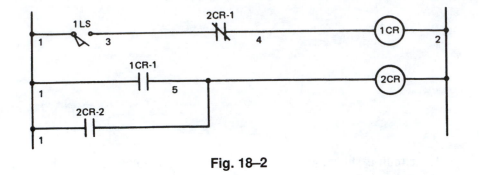

Fig. 18–2

9. Explain one problem in troubleshooting that can be helped with a good machine layout drawing.

10. In a large, complex circuit, what one item can be missed that leads to an open in a common line? Explain.

11. What are four items that can contribute to problems that arise with motor operation?

12. What consideration should be given when selecting the current range of a clamp-on ammeter you plan to use for troubleshooting?

13. What are some of the electrical equipment items that should be used when working on electrical circuits?

14. Name four important tools that a troubleshooter should have.

15. List three items that can be used to indicate component output.

16. What is the RMS value of ac voltage?

17. According to the National Electrical Code, for reasonable efficiency of operation, the voltage drop in branch circuits at the farthest outlet should be limited to what percent?

18. What problems can exist if excessive inrush current is present on motor starting?

19. List at least five items that should be on a safety check list for meters.

20. List at least two electronic troubleshooting hints.

21. What important problem can develop in a power circuit due to a loose connection?

22. In the circuit shown in Figure 18–3, explain how a ground is detected.

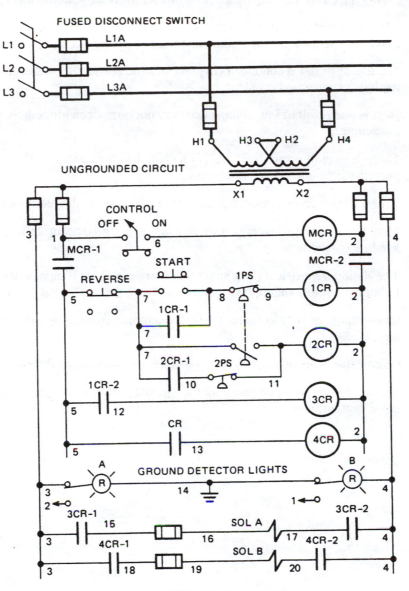

Fig. 18–3

TRUE AND FALSE STATEMENTS FOR CHAPTER 18

On the line provided, write TRUE if the statement is true or FALSE if the statement is false.

_____ 1. Always be certain that all power has been turned off, locked out, and tagged in any situation in which you must actually come in contact with the circuit or equipment.

_____ 2. The use of stranded conductors in place of solid conductors has in general greatly improved the loose connection problem.

_____ 3. Heat is one result of low voltage that may not be noticed immediately in the functioning of a machine.

_____ 4. There is overall economy in having a clean machine and a well-organized, well-executed preventative maintenance program.

_____ 5. In checking a given circuit, it is not important to know the resistance or impedance values.

_____ 6. Most meters called "average responding," do not give accurate RMS readings of the voltage in a pure sine wave.

_____ 7. The National Electrical Code states that for reasonable efficiency of operation, voltage drop in branch circuits should be limited to no more than 1% at the farthest outlet.

_____ 8. Overvoltage of 10% or more, applied to a motor on starting will not affect the inrush current.

_____ 9. Overheated parts in a circuit are a positive indication of problems.

_____ 10. Single phasing in a polyphase motor can generally be recognized by excessive noise and rapid heat buildup.

Chapter 19
Designing Control Systems
for Easy Maintenance

1. List several items that arc important in applying electrical control components in the assembly of a machine.

2. In Figure 19–1, cross-reference the relay contacts with the circuit line numbers.

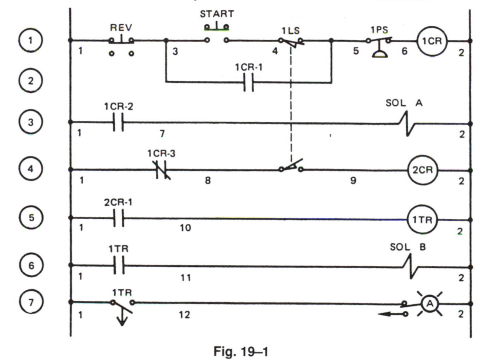

Fig. 19–1

3. Explain why it is important that the electrical components used in an electrical circuit be listed on a diagram and layout drawing. What would you include in the description of the components?

4. In Figure 19–2, a control panel is shown with components installed at random. Rearrange the components as you think they should be arranged. Add wiring channel and terminal blocks that you think would help in presenting a good panel design. Use graph paper to draw the panel design.

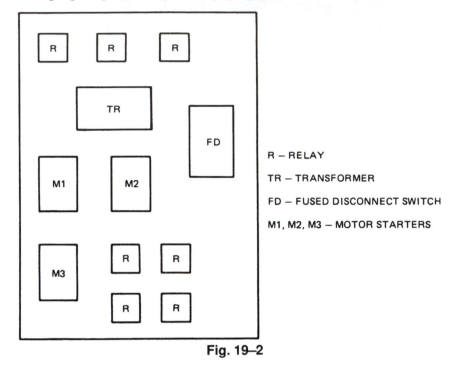

Fig. 19–2

5. What are the normal size of drawings showing electrical circuit and layout diagrams?

6. In Figure 19–3, explain the use of the broken line with an arrow and number.

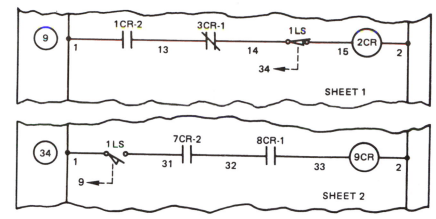

Fig. 19–3

7. What is the advantage in using a fuse in a solenoid circuit? What type fuse would you use?

8. What are some of the problems that exist in locating mechanical limit switches in a poor working environment?

9. In Figure 19–4, would you consider using the cam shown for operating a mechanical limit switch? If not, show the cam design that you would use. Use graph paper to draw your diagram.

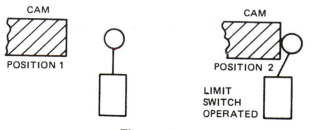

Figure 19–4

10. List two suggestions for good design in the mounting of push-button switches in an operating system.

11. What advantage is gained by supplying a reasonable description of a solenoid function on an elementary circuit diagram?

12. Are there any advantages in using plastic ties and wiring duct on a control panel as opposed to lacing the conductors? Explain.

13. What are two major areas to consider when designing for easy maintenance?

14. On a machine using multiple solenoids, why does it help to carry a reasonable description of their function?

15. What are some of the environmental problems found in using limit switches?

16. List some of the items to consider when installing motors, pressure switches, or temperature switches on a machine.

17. In Figure 19–5, what is the significance of the circuits shown with broken lines?

CONDUCTORS CARRYING LOAD CURRENT AT LINE VOLTAGE ARE DENOTED BY HEAVY LINES.

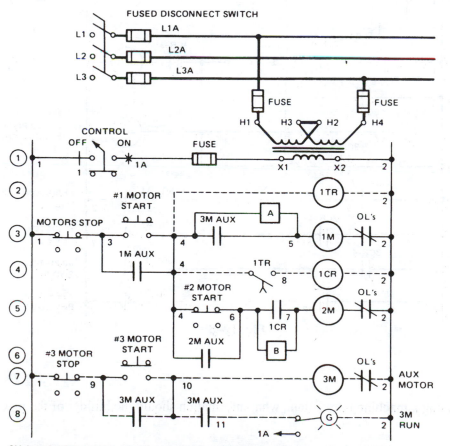

3M AUX PILOT MOTOR. OPTIONAL CIRCUITS SHOWN IN BROKEN LINE. TO REQUIRE 3M MOTOR STARTS BEFORE MAIN MOTORS 1M & 2M START, REMOVE JUMPER A. TO REQUIRE TIME DELAY BETWEEN 1M & 2M MOTORS STARTING, REMOVE JUMPER B.

Fig. 19–5

18. In Figure 19–6, what use is made of the numbers to the right and outside the vertical circuit line at the four relays: 1CR, 2CR, 3CR, and 4CR?

CONDUCTORS CARRYING LOAD CURRENT AT LINE VOLTAGE ARE DENOTED BY HEAVY LINES.

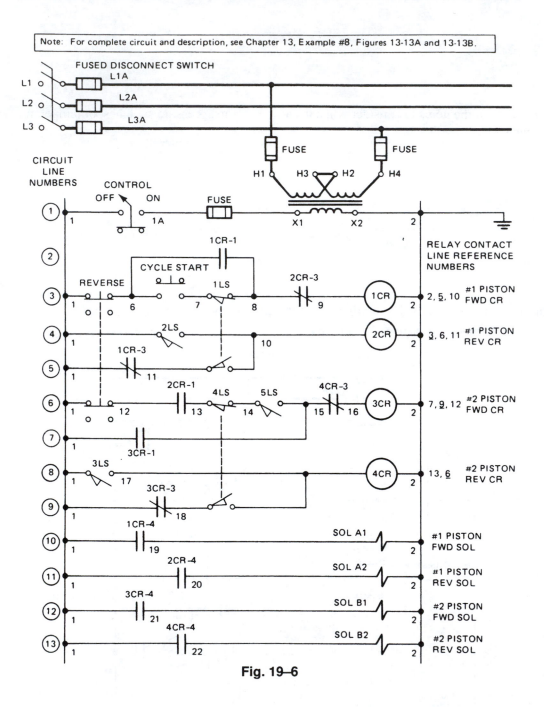

Fig. 19–6

19. Before receiving a machine into a plant, what information should the builder of the machine supply the user?

20. What is one advantage of using wiring duct on a control panel?

21. What care should be taken when assembling and wiring a push-button station?

22. In Figure 19–7, do you find any problem with the control panel layout? If so, explain.

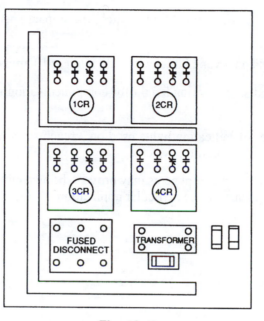

Fig. 19–7

TRUE AND FALSE STATEMENTS FOR CHAPTER 19

On the line provided, write TRUE if the statement is true or FALSE if the statement is false.

_____ 1. In a circuit diagram, N.C. relay contact reference numbers are so indicated by a bar below the reference line number.

_____ 2. A component listed on an elementary diagram or layout should be described so that if necessary a replacement can be obtained.

_____ 3. The usual drawing size used for circuit diagrams is 8" ↔ 10" or multiples thereof.

_____ 4. When wiring duct is not applicable, plastic ties can be used.

_____ 5. The use of fuses in a solenoid circuit is not economical under any circumstance.

_____ 6. In operating a mechanical limit switch, the impact speed should not exceed 400 feet per minute.

_____ 7. The contacts on a vane limit switch are rated at 10 amps make and 5 amps continuous.

_____ 8. An easy and safe working height for components mounted in a control cabinet is 2–5 1/2 feet.

_____ 9. The number of indicating lights used on a machine is usually directly proportional to the complexity of the control system.

_____ 10. The use of plug-in components may prove to be advisable if it is determined that a high rate of failure appears in one particular application of control.